Ford RANGER

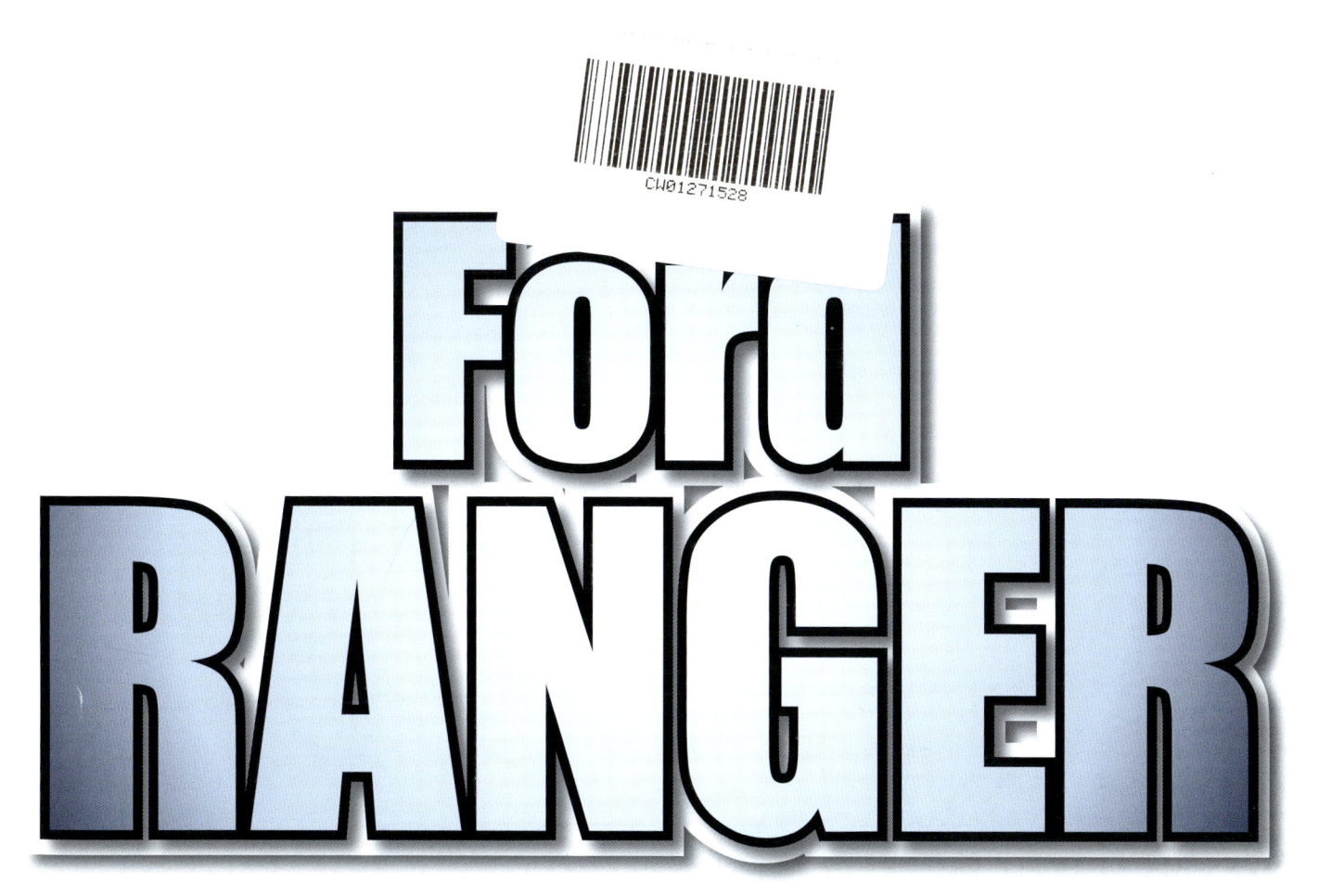

The Complete Illustrated History of America's Favorite Compact Pickup

plus bonus coverage of the Ford-badged Courier and the Ranger-based Bronco II

Paul McLaughlin

Enthusiast Books

Enthusiast Books

© 2016 Paul McLaughlin

All rights reserved. No part of this work may be reproduced or used in any form by any means. . . graphic, electronic, or mechanical, including photocopying, recording, taping, or any other information storage and retrieval system. . . without written permission of the publisher.

The information in this book is true and complete to the best of our knowledge. All recommendations are made without any guarantee on the part of the author or Publisher, who also disclaim any liability incurred in connection with the use of this data or specific details.

We acknowledge that certain words, such as model names and designations, mentioned herein are the property of the trademark holder. We use them for purposes of identification only. This is not an official publication.

To contact us or request a catalog, go to www.enthusiastbooks.com or call us at 1-715-381-9755.

Library of Congress Control Number: 1-3121072921

ISBN-13: 978-1-58388-333-4
ISBN-10: 1-58388-333-9

Printed in USA

Contents

Dedication/Credits	4
Chapter 1: Pre-Ranger Days	5
Chapter 2: The Ranger Debuts As America's Compact Truck Built Ford Tough	14
Chapter 3: The Best Built American Trucks Are Built Ford Tough in 1984-1985	24
Chapter 4: 1986-1988 More Ranger Models and Introducing the SuperCab	32
Chapter 5: A New Look To America's Best Selling Compact Truck	44
Chapter 6: 1993-1997 The New Look Rangers Make Their Debut	56
Chapter 7: 1998-2000 Making A Good Truck Even Better And More Appealing	70
Chapter 8: 2001-2003 More Power, More Comfort, and More Changes For The Ranger	80
Chapter 9: America Still Loves The Ranger 2004-2005	90
Chapter 10: 2006-2009 Bolder And More Aggressive Looking Rangers	100
Chapter 11: 2010-2012 Heading For The Finish Line	108
Chapter 12: 1984-1990 Bronco II, Ford's Ranger Based SUV	118

Dedication

I would like to dedicate this book to all those Ford Motor Company employees who were involved with the Ranger program here in North America. From the men and women who bolted them together on the assembly lines to the designers and engineers who helped to develop them from the late 1970s until the last one left the Twin Cities Assembly Plant in December 2011.

Paul G. McLaughlin
October 2015

Credits

I would like to thank all the people who supported me during this project as well as the people who have supported me during my other book projects down through the years.

Paul C. and Melanie McLaughlin
S. Bernice McLaughlin
Robert T. Lucero, DDS
Ken Campbell
Matt and Jessica Kerschen
Amy and Lonney Kimmel
Paul G. and Frances McLaughlin
Sam and Julia Madrid
Richard McLaughlin
John McLaughlin
Bill Stroppe
George and Claire Hinds
Patricia Curran
Austin Kerschen

P.J. McLaughlin
Louise and Francis Judd
Al and Jean Clarke
Rob Campbell
Mike and Charlene Benavidez
Fran and Paul McDonald
Joe Rose
Ruth and Jim Woodard
Joseph McLaughlin
Dylan Frautschi
Mustang Cafe Breakfast Group
Joan McLaughlin
Michael Gonzales

This Promotional Courier postcard was released by Ford about the same time that the new Couriers hit the streets.

Chapter 1: **Pre-Ranger Days**

Prior to the early 1970's the Big Three American Automobile Manufacturers had little, if any, interest in producing or selling trucks in the compact truck segment of the market; a market segment that didn't really start to develop until the mid to late 1960s.

The only notable American company that had built a "mini truck" prior to this time was Crosley. Powell Crosley introduced his small pickup truck in 1939 as a 1940 model. There was also a 1941 Crosley pickup model but production of these vehicles ceased when the United States entered World War II.

Powell Crosley started producing his Crosley pickup again for the 1947 model year and continued to produce his small pickups into the 1952 model year when his company was sold to the General Tire and Rubber Company.

In 1950 Powell decided to expand his "mini-truck" lineup by introducing a new model called the "Farm-O-Road" pickup. This "Farm-O-Road" pickup looked like a downsized Jeep. After Crosley went out of business one of the companies that bought some of the assets of the old Crosley Company was a company called Crofton Marine Engineering and this company came out with their own updated version of the "Farm-O-Road" pickup which they called the "Crofton Bug."Crofton stopped producing their "Crofton Bug" around 1962.

Both companies, Crosley and Crofton, were not very successful in convincing American businessmen to buy their small pickup truck models. At best their annual sales of these "mini-trucks" probably only amounted to a few hundred units.

At about the same time Crofton was introducing their "Crofton Bug" Datsun began exporting their mini-pickup to the USA. In 1959 Datsun sold less than a dozen of these vehicles in the United States

5

Here is another early model Ford Courier promotional postcard from the early 1970s.

and over the next few years Datsun sold a grand total of 300 or so units, with most of them being sold to people living in California.

Datsun kept improving their little pickup trucks to make them more appealing to the American buyer, improving them so much that by 1965 Datsun truck sales in the United States numbered around 8000 units. A pretty impressive amount when you consider how few they sold six years earlier.

That sales number prompted Toyota to bring their own mini-pickup to these shores as well in 1965. These trucks were not as popular as the Datsun models for between 1965 and 1967 Toyota only sold about 4200 of these trucks in the United States.

Like Datsun did before them, Toyota made improvements to their mini-trucks and though annual sales increased, American buyers still preferred the Datsun over the Toyota by a wide margin. Between 1965 and the end of the 1969 model sales year Datsun sold some 66,000 mini-pickup trucks in spite of the competition from Toyota.

That 66,000 unit number caused Detroit to stand up and take notice that this trend to smaller, more compact, pickup trucks was not a passing fad, and maybe they should consider entering this market segment as well and not leave this end of the market to be dominated by the Japanese companies.

Datsun was the undisputed leader at this end of the market and there was no doubt about that leadership position when Datsun sold almost 67,000 mini-pickups in 1971, the last model year before the American manufacturers decided to take this fight to the Japanese companies for their fair share of sales.

Back in the 1960s it took a minimum of three full years for the car and truck manufacturers to bring along a car or truck from the initial design stage to a fully functional, ready for sale, marketable vehicle.

The Big Three decided to follow a different path this time because they all felt that they couldn't leave this growing market segment for three or four more years to the Japanese companies themselves. Rather than commit the time and effort to develop their own new compact pickup trucks they decided to join forces with some Japanese manufacturers to bring their own small pickup trucks to market a lot sooner.

Chevrolet decided to partner with Isuzu Motors LTD by buying a little over thirty percent of the company. The marriage of these two companies produced the Chevrolet LUV mini-pickup that was available for sale at Chevrolet dealerships across the USA.

The Ford Motor Company wasn't about to be left behind in this rush to enter the compact truck market segment so they partnered with Toyo Kogyo LTD, the same company that produced the Mazda

FORD's COURIER... WINS BAJA 500

This promotional postcard shows that Ford was already promoting the Courier as one tough racing truck.

mini-pickup. The Ford version that came out of this collaboration was called the Courier.

Chrysler's Dodge Division also decided to partner with a Japanese company to produce a mini-pickup truck but they weren't ready to jump into the fray like Chevrolet and Ford until later in the decade. Their partner in this endeavor was Mitsubishi and the mini-truck they produced was the Dodge D-50.

Mazda built the Ford Courier in their plants and shipped it over here where it was introduced at Ford dealerships on the West Coast in March of 1972. The Ford Courier pickup looked pretty similar to the Mazda Sport Truck. The only real differences between the two was that the Courier used a Ford styled grille that looked similar to the grilles used on Ford's full size pickup trucks along with single headlights. The Mazda truck used a different tri-bar styled grille and dual headlights per side.

Under the hood the Mazda version used an inline SOHC 4-cylinder engine that displaced 96.8 cubic inches or 1600 cubic centimeters while the Ford Courier used an inline SOHC 4-cylinder engine that displaced 110 cubic inches or 1800 cubic centimeters.

The Mazda engine was rated at 70 horsepower at 5000 rpm while the Ford Courier's engine was rated at 74 horsepower at the same 5000 rpm level. Both trucks were fitted with drum brakes, manual steering, and all-synchro 4 spped manual transmissions.

The Mazda had a load capacity of 1000 pounds while the Courier's maximum load capacity was listed at 1100 pounds. Curb weights for both trucks were only about forty pounds apart with the Courier weighing in at 2510 pounds and the Mazda Sport Truck at 2470 pounds.

When it came to mileage figures the Mazda scored higher on the highway average figure of 25 mpg compared to the 23 mpg scored by the Courier. However on the city average the Courier scored higher than the Mazda with an average of 24 mpg versus the Mazda's figure of 19 mpg.

Base price of this new Ford Courier compact pickup truck was set at $2222.00, a nice round number for a truck of this size that was rated at a half ton: the perfect size for a business that needed a smaller truck with better fuel economy numbers.

This new size Ford truck with its mid model year introduction didn't do too badly on the sales charts because Ford sold almost 27,000 units in a shortened model year span of eight months. That was considered a pretty good number for such a new model wearing a Ford label.

Ford made only minor changes for their 1973 Courier models sometime during the model year. They also added an extra cost automatic transmission to the Courier option list for those buyers who didn't want to deal with a clutch and shifting a manual transmission. Ford also tried to make the 1973 Courier more appealing to a wider range of buyers by offering an exterior dress-up package, air conditioning, tinted glass, deluxe exterior mirrors, and more items for an

Here we see an original copy of the 1976 Ford promotional brochure that was given out to potential Courier customers when they visited Ford dealer showrooms.

This 1977 Ford Courier promotional catalog shows an XLT trimmed model in the foreground and a "Free-Wheelin'" Courier in the background.

owner to personalize his vehicle.

The Ford Courier stayed basically the same with no major changes for the model years of 1974 and 1975. Sales totals for the Courier in 1974 amounted to 44,491 and in 1975 sales were even higher when 56,073 units left Ford dealerships.

If there were any real complaints having to do with this Courier compact truck, those complaints might have had something to do with the claim that Ford said you could fit three adults into the cab comfortably. That might be all well and good if those passengers were thin and were less than six feet tall. But if you were a normal-sized American driver with two other passengers in the Courier cab you might have felt a little cramped.

For 1976 Ford introduced a three inch longer redesigned cab for the Courier which allowed the seat to move further back in the cab making the cab a little more comfortable for taller drivers and passengers. The 1976 Courier was also updated with a new grille treatment, fancier looking vinyl interior, and more convenience items added to the Courier option list.

Another change on the 1976 Ford Couriers was the addition of a new 5-speed manual transmission option. When equipped with this 5-speed manual transmission, a Courier compact pickup truck had the potential to hit really high mpg figures of 36 mpg on the highway and 23 mpg in city driving situations.

For those buyers who didn't need a Courier with a maximum payload rating of 1400 pounds but would rather have a more comfortable riding Courier pickup, Ford offered an option they called "The Soft Ride Option" for the Courier. This option utilized softer rated springs and recalibrated shock absorbers to achieve the desired affect.

For buyers who wanted a fancier looking Courier pickup truck, Ford offered them "A Convenience and Decor Option" that consisted of color-keyed, 14-ounce cut pile carpeting, engine compartment light, deluxe simulated wood grained interior accent trim pieces, bright plated bumper guards, body side moldings, and full wheel covers.

In addition to pickup trucks for the first time in Courier truck history Ford offered customers a "Cab and Chassis Option" model that they could add their own body to.

Even with all these changes that were designed to make the Courier more appealing to a wider range of buyers, the 1976 Ford Courier production total dropped to 51,400 units, a loss of some 4000 units.

Early model 1977 Ford Couriers looked pretty much the same as the 1976 models that preceded them, but in the spring of 1977 Ford introduced a restyled Courier with a new grille, new front bumper, restyled hood and front fenders, and other exterior body changes that gave these "1977 1/2" Couriers a more modern look.

The 1977 1/2 Courier was also now available with two different wheelbases and pickup box lengths. The regular Courier sat on a wheelbase of 106.9 inches with a six-foot long bed. For those buyers who wanted a longer truck bed, Ford offered a seven-foot bed for the Courier and these trucks featured a 112.8 inch wheelbase. The overall length of a Courier with a six-foot pickup box was 177.9 inches while the overall length of a Courier with a seven-foot bed was 189.4 inches.

Another new feature found on these 1977 1/2 Courier trucks was the addition of power front disc brakes replacing the old manual front brake drums used on earlier Couriers.

1977 1/2 Courier Exterior Colors
Dark Red
Bright Orange
Yellow
White
Green
Blue

Optional Extra Cost Colors
Bronze Metallic
Silver Metallic
Blue Metallic

Courier buyers could now choose between a regular model Courier or spend a little more money and move up to a fancier looking Courier XLT model. This deluxe XLT version came standard with a bright plated grille, bright windshield and rear window molding, drip rails, and wheel lip moldings along with "XLT" nameplates, deluxe wheel covers, and dual accent pin striping.

The Courier XLT interior featured soft supple vinyl seat and door panel trim, wood-grained vinyl covered shift knob, cut pile carpeting, day/night mirror, and lights in the ashtray, glove box, and

1978 Ford Courier Options with Prices

2.3 Liter Inline 4-Cylinder Engine	$162.20
5-speed Manual with Overdrive Transmission	$154.70
3 Speed Automatic Transmission	$334.30
Air Conditioner	$461.60
Rear Step Bumper	$82.30
California Emissions Package	$99.80
Cold Weather Group (Rear Window Defroster, Heavy Duty Battery, High Output Heater)	$118.80
Fog Lamps (Free Wheeling Package A)	$81.10
Tinted Glass	$26.50
Western Style Mirrors	$46.70
Dual Accent Paint Striping:	
(Without XLT Trim Package)	$165.30
(With XLT Trim Package)	$120.40
7 Foot Pickup Box	
(Includes 17.5 Gallon Fuel Tank)	$206.60
Protection Exterior Group	$70.70
Pushbutton AM Radio	$75.70
AM/FM Monaural Radio	$143.50
Soft Ride Springs	$33.70
Free Wheeling Package A	$212.10
Free Wheeling Package B:	
(Without XLT Trim)	$746.00
(With XLT Trim)	$661.30
XLT Trim Group	$333.00
Sliding Rear Window	$71.10
(5) 195 Rx 14 inch Steel-belted Radial Tires	$128.25
(5) E70x14 inch Raised White Letter Tires with 4 Cast Aluminum Wheels	$316.90
(No Charge with Free Wheeling Package B)	

engine compartment. Other XLT interior features included a glove box lock, a wood grain appliqué on the steering wheel hub and on the dashboard, and water temperature and ammeter gauges.

In the late 1970s Ford was making an all out push to make their products look more appealing to young people and one of the vehicles they "jazzed" up was the Courier with two optional "Free Wheeling Packages" for their 1977 1/2 models.

"Free Wheeling Package A" came with a matte-black, painted hood panel; a black, painted GT Roll Bar mounted in the bed; and a black, painted push bar on the front of the vehicle to protect the front bumper and grille from damage.

"Free Wheeling Package B" came with all that equipment plus fog lamps mounted on the push bar, cast aluminum custom wheels, raised-white-letter 70 Series tires, and a striking two color body side tape stripe arrangement.

And finally for those Courier buyers who wanted a little more power Ford offered a 2.3 Liter inline 4-cylinder engine as an optional power upgrade.

For 1978 the base price of a Courier was set at $3900.00 and for that amount of money the buyer received a standard model Courier with a restyled grille, which now housed a set of parking lamps along with a restyled front bumper.

If a standard model Courier wasn't your cup of tea in 1978 you could still order a Courier XLT, or a Courier with both "Free Wheeling Packages" offered before with the 1977 1/2 models. Evidently buyers liked what they saw with the 1978 Couriers because the production total for these compact, or mini-pickups, topped the 70,500 unit mark.

Ford's 1979 Couriers looked pretty much the same as their 1978 models. Once again buyers could choose between a regular model Courier, and the deluxe XLT trimmed Courier. And for sporty young buyers there were still Couriers with the "Free Wheeling Package A" or the "Free Wheeling Package B" options.

Another "youth-oriented" package that was offered for the 1979 Couriers was a new option that Ford called their "Sports Group Package". This new "Sports Group Package" consisted of a black-colored sport steering wheel, bucket seats trimmed in black vinyl with black and white plaid cloth inserts, wood toned instrument panel accent trim, black carpeting, black door panels, and a temperature and ammeter gauges.

A new 2.0 Liter Overhead Cam inline 4-cylinder

engine was the standard powerplant for the Courier now with the 2.3 Liter inline 4-cylinder engine as a power upgrade option for the buyer who wanted more power.

By the end of the 1979 model year Ford had sold some 76,880 of them making this the best year in Courier sales since they were introduced back in March of 1972.

Things stayed pretty much the same for the Ford Courier in 1980 and 1981. Though there were no major changes of note for these models their sales were still pretty good. Sales for the 1980 Courier were around 78,400 units and sales for the 1981 models were stood at 66,155 units.

Ford referred to their 1982 Courier truck when it was introduced in late 1981 as their "Tough 1982 Courier" which, as we now know, was entering its last model year. After a ten-year run with the Mazda built, Ford-badged truck, they were about to release their own Ford built truck to compete in the compact truck market.

This new Ford truck was called the Ranger and it made its official debut in March of 1982, almost ten years after the Courier came on the scene. With this new model on Ford dealer lots to sell, it must have been hard for Ford salesmen to sell the old truck when the buyer could see and buy a new Ford truck.

The Courier on the cover of this 1982 catalog is trimmed out like an XLT model in a bright blue color.

Have you ever heard of the "Coors Courier" before? This customized truck was built to be shown at custom car shows around the country. This promotional postcard was part of the "Ford Motorsport Series" that was released in the early 1980s.

Chapter 2:
The Ranger Debuts As America's Compact Truck Built Ford Tough

As we noted in the last chapter, in order to get into the mini-truck or compact truck market segment in the early 1970s Ford partnered with Mazda to produce a very popular truck called the Courier.

That truck made its debut in 1972 and over the course of the next ten years, Ford used the time to develop their own truck, which would turn out to be more popular than the earlier Courier. Ford called this truck the Ranger and they officially introduced it on March 12, 1982 as a new 1983 model.

Before Ford actually showed this Ranger "in-the-metal" so to speak, Ford had mailed out promotional materials about their new Ranger to about two million potential customers. And some of those potential customers showed up at their local Ford dealership on March 12, 1982 to see and learn about this new smaller-sized Ford pickup truck.

Ford promoted these new compact trucks as being "Built Like The Big Ones, Saves Like The Little Ones". A claim that was pretty true because these new Rangers looked like scaled down F Series pickup trucks.

Besides promoting these new "little" Ford trucks to the public Ford also built up interest in these new Rangers to the people who were tasked to sell them to the public.

The marketing and training people at Ford promoted these new Ranger trucks through company periodicals like their "Dealer World" and "Salesforce" magazines and newsletters.

Some of you are old enough to remember the "Lone Ranger" western-themed television show of the 1950s. One of the major stars of that program was the late Clayton Moore who played the leading role as the "Lone Ranger". For the cover of the January/February 1982 issue of Ford's "Dealer World" magazine the editors pictured a new Ranger pickup with the "lone Ranger" in a clever photograph with the title "Unmasking the Ranger" and inside the magazine there was a two-page article introducing

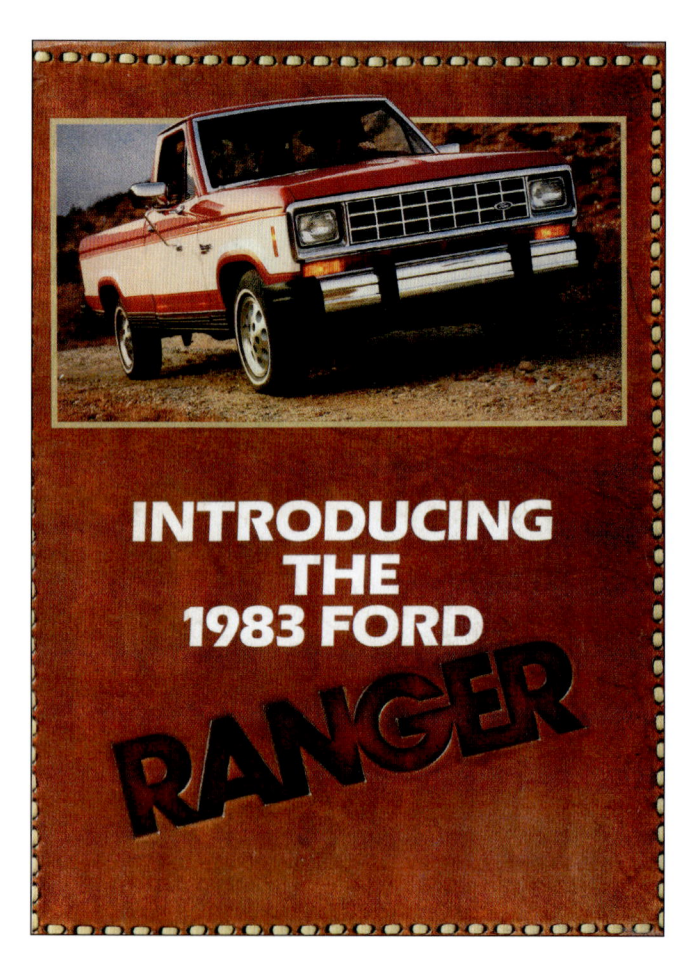

This tooled leather looking cover appeared on the promotional showroom catalog introducing the new 1983 Ford Ranger in the spring of 1982.

the Ranger to Ford salesmen.

In an issue of Ford's "Salesforce" newsletter released around the same time, the editorial team spotlighted the new Ranger with an article and the headline "A Total Quality Commitment the 1983 Ranger Introduced".

This article talked about how Ford had modernized their Louisville Assembly Plant to build this new Ranger. The writers also talked about the fact that the body shop part of the plant could produce one truck body every thirty six seconds and it could do that through the use of some automated equipment and a plant full of dedicated employees who were committed to producing a high quality product.

At that time the Ford Motor Company promoted this truck as being…"American-built, with an excellent ride and more room than the imports with big truck features and the tough Ford heritage".

Another promotional slogan that Ford used then was "Born and bred in the pioneering spirit the Ranger caters to the free-wheeling lifestyles of hard working, hard-playing, and hard-driving Americans". Compared to the Courier, the new Ranger is wider and sits higher with a roomier cab than the old truck.

This new Ranger featured double-wall construction in the roof, hood area, pickup box sides, and tailgate to give the truck a very solid quality feel. The Ranger's cab and pickup box were installed over a tough truck style ladder type frame and on the front side of that frame was a "Twin I Beam style suspension system just like the one you would find under a full size Ford 4x2 pickup.

There were two wheelbase lengths used for this new Ranger. The wheelbase for trucks fitted with 6-foot long pickup boxes was 108 inches long while Rangers equipped with the optional 7-foot long pickup boxes had a wheelbase of 114 inches.

Under the hood, Ford installed a new inline Overhead Cam 4-cylinder engine that displaced 122 cubic inches, or 2.0 Liters and had a maximum horsepower rating of 73 at 4000 rpm as the standard engine for these trucks. This engine was combined with a 4-speed manual transmission and with this combination these new Rangers carried an estimated 39 mpg highway rating along with an estimated city average of 27 mpg. High fuel economy numbers for any truck of any size and a major consideration for buyers who were looking to purchase a dependable and very economical truck in 1982 and 1983.

For those truck buyers looking for a little more power Ford offered a larger 2.3 Liter Overhead Cam inline 4-cylinder engine as an extra cost option in all states except California. For Rangers sold new in California this 2.3 Liter Overhead Cam engine was the standard engine. That 2.3 Liter engine carried a maximum horsepower rating of 79 at 3800 rpm.

Specification Sheet

	1982 Courier	1983 Ranger
Overall Height	61.5 inches	64.02 inches (6 Foot Box)
		63.98 inches (7 Foot Box)
Overall Width	63.0 inches	66.9 inches
Wheelbase	106.9 inches (6 Foot Box)	107.9 inches (6 Foot Box)
	112.8 inches (7 Foot Box)	113.9 inches (7 Foot Box)
Overall Length	177.9 inches (6 Foot Box)	175.6 inches (6 Foot Box)
	189.4 inches (7 Foot Box)	187.6 inches (7 Foot Box)
Bed Width at Floor	61.4 inches (6 and 7 Foot Boxes)	54.3 inches (6 and 7 Foot Boxes)
Base Curb Weight	2680 Pounds (7 Foot Box)	2526 Pounds (6 Foot Box)
		2559 Pounds (7 Foot Box)

> **1983 Ranger Exterior Color Choices**
>
> Raven Black
> Silver Metallic
> Medium Grey Metallic
> Candyapple Red
> Midnight Blue Metallic
> Medium Blue Metallic
> Bright Blue
> Dark Spruce Metallic
> Dark Brown Metallic
> Medium Yellow
> Bittersweet Glow (Extra Cost Option)
> Bright Bittersweet (4x4 Models)
> Wimbledon White
> Desert Tan
> Light Desert Tan
>
> **1983 Ranger Interior Colors**
>
> Black
> Dark Blue
> Red
> Tan

In the 1983 model year the Ranger was available in four trim levels. These trim levels, or models, were the Ranger, Ranger XL, Ranger XLT, and Ranger XLS.

The base model Ranger came equipped with a black-colored front bumper, windshield wiper arms, front spoiler, exterior mirrors, a two-spoke steering wheel, and a black- colored vinyl coated rubber floor mat. Other standard equipment found on this model included a chrome plated grille, a bright windshield molding, a removable tailgate; argent (silver) colored, styled-steel wheels; an AM radio, a Day/Night rear mirror, color-keyed vinyl covered door panels, a vinyl color-keyed seat cover, and color-keyed vinyl covered sun visors.

The Ranger XL came standard with all that equipment or upgraded equipment which included a chrome plated front bumper, chrome plated grille, rear window molding with bright insert, a black-colored front spoiler and black windshield wiper arms, bright plated windshield molding; argent-colored, styled-steel wheels with bright plated trim rings and hub caps; an AM radio, wood toned instrument cluster trim, cigarette lighter, and left hand and right hand

Another 1983 Ford Ranger showroom catalog showing the Ford Ranger as one of Ford's "Built Ford Tough Trucks". A similar catalog was used to introduce the Ranger diesel trucks.

Still another Ford Ranger showroom catalog introducing both the new V-6 engine option and the diesel engine option.

courtesy light switches.

Other upgraded Ranger XL pieces included a color-keyed cloth headliner, color-keyed vinyl covered door panels, color-keyed vinyl covered bench seat, color-keyed seat belts, aluminum scuff plates, bright wheel lip moldings, and color-keyed vinyl coated rubber floor mats.

For those who wanted a really deluxe looking Ranger the Ranger XLT was the model to buy this year. The XLT came with the following standard equipment this year in addition to or different from the standard equipment that came on the lower tiered models; a chrome plated front bumper with black-colored end caps, chrome plated grille, bright plated windshield molding, rear window molding with bright insert trim, dual accent body striping, full length black vinyl lower body side molding, black-colored 4-spoke steering wheel, Day/Night mirror, contoured cloth and vinyl combination bench seat cover, color-keyed seat belts, and color-keyed carpeting.

Other standard equipment found on the Ranger XLT included an AM radio, color-keyed cloth headliner, wood toned instrument cluster trim pieces; argent-colored, styled-steel wheels with beauty rings and bright hub caps; and aluminum scuff plates.

The last 1983 Ranger model was the model that Ford called their "Young At Heart" Ranger, the Ranger XLS with the "S" standing for "Sassy" and or "Sporty".

This model was designed for the "fun loving" Ranger buyer who wanted a distinctive looking model for the young or the young at heart.

The Ranger XLS came standard with distinctive looking black out trim including a black-colored grille, black-colored front and rear bumpers, front spoiler, windshield molding, exterior mirrors, and windshield wiper arms.

Inside the Ranger XLS featured some luxury touches like those seen in XLT models. Deluxe touches like a soft wrap steering wheel, cloth and vinyl combination seat covers on reclining bucket seats, a cargo light, bronze-tinted gauge cluster trim, full color-keyed carpeting, color-keyed cloth headliner, and a dome light.

On the outside the Ranger XLS came with unique "XLS" two color tape striping and argent-colored, styled-steel wheels with bright plated beauty rings, and black-colored hub caps.

Ford offered their Ranger pickups in a variety of

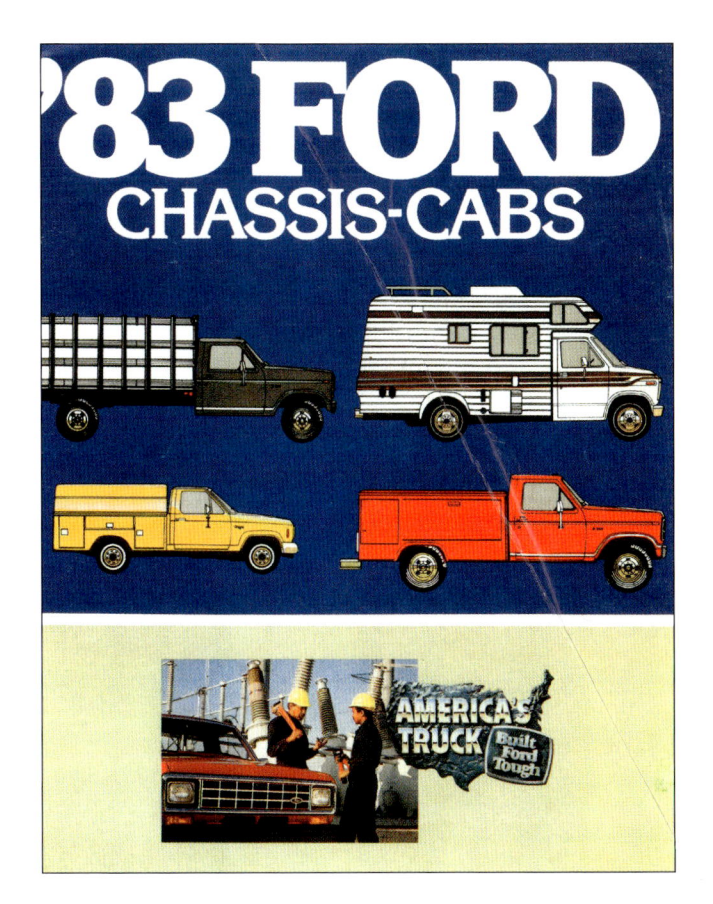

The cover of this Ford Chassis Cab brochure shows a drawing of a non-pickup Ranger truck equipped with a utility body.

exterior colors. In addition to single color choices, Ford offered the Ranger buyer his or her choice three unique two-tone, extra-cost exterior paint combinations.

The first option Ford called the "Regular Tu-Tone Paint Option" which had the primary color applied to the top, hood, and upper body areas front to back. An accent color was then applied to the lower body side and tailgate areas along with a two-color tape accent stripe.

Option number two was called the "Deluxe Tu-Tone Paint Option" which consisted of the primary color being applied to the top, hood, and upper and lower body areas along with upper and lower two color tape accent striping. The accent color on this option was applied to the mid-body section front to back and the tailgate areas.

And the last Ranger two-tone paint option was called the "Special Tu-Tone Paint Option". This version had the primary color applied to the top, hood, and the upper body areas from the doors forward and the lower body areas front to back.

1983 Ford Ranger Options

Interior Appearance Package (Includes color-keyed cloth headliner, upper door moldings, cowl trim panels, rear window moldings, and aluminum scuff plates)
Regular Tu-Tone Paint Option
Deluxe Tu-Tone Paint Option
Special Tu-Tone Paint Option
Air Conditioning
Black-Colored Rear Step Bumper
AM Radio
AM/FM Monaural Radio
AM/FM Stereo Radio
AM/FM Stereo Radio with Cassette Player
Convenience Group
(Includes dual electric horns, interval windshield wipers, Passenger side visor mounted vanity mirror, and cigarette lighter)
Tinted Glass
Light Group
(Includes ashtray light, cargo light, passenger door courtesy light switch, glove box light, and "headlights on" buzzer)
Power Steering
Low Mount Western Style Mirrors
Overhead Console
Tilt Steering Wheel
Sliding Rear Window
Cargo Tie Down Hooks
Pivoting Vent Windows (Standard on XLT)
Speed Control
Reclining Bucket Seats
Knitted Vinyl Seat Cover
Cloth and Vinyl Seat Cover Combination (Included with XLT Trim)
Extra Payload Package
5-Speed Manual Overdrive Transmission
Engine Cooling Package
2.2 Liter Diesel Engine (Not available on 4x4 Rangers)
Snow Plow Special Package
Automatic Transmission (Only available on 4x2 models with 2.3 Liter engine)
Heavy Duty Battery
Heavy Duty Air Cleaner
Camper Package
Gauge Package (Includes ammeter, oil pressure, temperature, and trip odometer, standard on XLS models)
Auxiliary Fuel Tank (13 Gallons)
Engine Block Heater
Heavy Duty Shocks
Traction-Lok Rear Axle
4x4 Limited Slip Front Axle
Tow Hooks
Trailer Towing Package
Heavy Duty Front Suspension
Handling Package
4x4 Skid Plates
High Altitude Emissions System
California Emission System
Security Lock Group
Exterior Protection Group
Cast Aluminum Wheels
White-Colored Sport Wheels
Deluxe Wheel Trim

The accent color was then applied to the mid-body section, inside the pickup box as well as the upper and mid body sides of the pickup box, rear area of the roof, and the tailgate. Also included in this option were upper and lower two-color tape stripes.

As you can readily see, Ford made sure the Ranger buyer in 1983 had a lot of equipment and packages available to tailor-make a vehicle to meet anyone's needs or tastes.

When Ford introduced their new 1983 Ranger pickups in March of 1982 they brought a very salable compact vehicle to the market, but the company was planning on offering even more to the Ranger buyer as the model year progressed.

When the 1983 model year started, Ford offered the Ranger buyer two 4-cylinder engine options, two transmissions, and two pickup box lengths in a 4x2 configuration.

Less than six months after the first Ford Rangers hit the streets the company offered buyers a 4x4 option layout for the Ranger. And soon thereafter, Off Road Magazine picked the Ranger 4x4 truck as their "4x4 Truck of the Year". Quite an honor for a truck that was so new to the market.

About the same time the 4x4 option was released for the Ranger Ford upped the economy ante by adding a 2.2 Liter 4-cylinder diesel engine to the Ranger Option List.

This diesel engine option was only offered on 4x2 Ranger trucks.

Another new Ranger option introduced at about the same time for 2.0 Liter and 2.3 Liter 4-cylinder Rangers was a new 5-Speed Manual/Overdrive transmission. This transmission consisted of 4 fully synchronized forward gears, an overdrive gear, and a reverse gear. When equipped with this new transmission the Ranger buyer could expect even better fuel mileage figures.

As the model year was coming to a close, Ford released another engine option for the Ranger in the guise of a 2.8 Liter V-6 engine that was electronically controlled by Ford's excellent EEC-IV Electronic Engine Control System. This engine carried a maximum horsepower rating of 115 at 4600 rpm and a maximum torque rating of 150 Lbs/ft at a low 2600 rpm. This was more than enough power and torque to move a Ranger out quite smartly.

So far we have concentrated our attention on Ranger pickup trucks, but Ford also offered the Ranger without a pickup box. Ford called these non-pickup trucks their "Chassis and Cab" models. By offering this option on long wheelbase 4x2 Rangers, a buyer could install an aftermarket body on a Ranger chassis to meet just about any need a commercial truck buyer might have.

Before 1983 came to a close Ford introduced a new vehicle that was based on the short Ranger platform. This new vehicle was a downsized sport utility vehicle that Ford called their Bronco II. This Bronco II vehicle was listed as a 1984 model and it was available in three different trim levels as the base model Bronco II, the deluxe version called the Bronco XLT, and the sporty Bronco XLS.

All these Bronco II models came standard with four-wheel drive and Ford's 2.8 Liter V-6 engine. This was the perfect size, a small sport utility vehicle for buyers who wanted a smaller, more economical, and sporty themed vehicle that they could take off road whenever they wanted to go trail blazing.

At the end of the 1983 model year Ford had two compact sized vehicles that they could be proud off. In the case of the Ranger it was a better vehicle than the Courier it replaced and the Bronco II gave 4x4 buyers and off-road enthusiasts the chance to buy a smaller sport utility vehicle that was more economical to operate and more nimble than it's bigger brother.

Here we see a photo of a racing Ranger pickup gracing the cover of a Service Life magazine.

Here is an example of an "artsy" Ford magazine advertisement introducing the new 1983 Ford Ranger pickup truck.

As you can see here this Ford magazine ad promotes the new V-6 engine powered Ranger and compares the Ranger to Chevrolet's S-10 pickup truck.

18

Another magazine ad comparing the V-6 Ford Ranger to the Chevrolet S-10 pickup. It also talks about the new diesel engine option available in the Ranger pickup.

This promotional Ford postcard shows a Ranger XLT truck decked out in a two-tone paint scheme and other dress up items.

This copy of the Ford Salesforce newspaper talked about the new 1983 Ranger trucks built at the Louisville Assembly Plant and how Ford salesmen should try to sell them.

19

"The Best Built American Trucks are Built Ford Tough" including the new for 1984 Bronco II, the Ford Ranger, and Ford's F Series trucks.

Chapter 3:
The Best Built American Trucks Are Built Ford Tough in 1984-1985

Inside the 1984 showroom catalog, Donald E. Peterson, then President of the Ford Motor Company introduced the new Ford Ranger and other Ford products of that year with the following quotation.

..."At Ford, quality is our top priority. Nothing ranks higher in the design, engineering, manufacture, sale, and service of our cars and trucks. We know that our jobs and the success of Ford Motor Company depend on building high quality vehicles that meet your needs and expectations".

In that same vein, 1984 was the year when the Ford Motor Company instituted their promotional "Quality is Job 1" campaign.

As far as the Ranger went Ford felt pretty good about this compact truck which was entering its second model year of production. After the 1983 Ranger had been in production for three months or so, Ford conducted an owner survey of how these new owners felt about their Ranger trucks. A large percentage of those owners who responded to the survey told Ford they were really satisfied with their Ranger trucks. And that made Ford executives happy because it meant that their customers had a good feeling about the quality of vehicles that the company was building.

Once again the Ranger lineup consisted of the same four models, or trim levels that were available in the previous model year. They were the Ranger (base model), the Ranger XL (deluxe model), the Ranger XLT (super deluxe model), and the Ranger XLS (sporty model).

The power train options were also carried over with the 2.0 Liter OHC inline 4-cylinder engine, the 2.2 Liter inline 4-cylinder diesel engine, the 2.3 Liter OHC inline 4-cylinder engine, and the 2.8 Liter V-6. Transmission choices were still the 4-speed manual, the 5-speed manual with overdrive, and a 3-speed automatic.

Ford expanded their special Ranger optional packages this year to help buyers tailor-make their vehicles to better meet their expectations.

One of those options that was upgraded this year was Ford's "Snow Plow Special Package" which as the name suggested was perfect for the Ranger buyer who lived in snow belt areas and needed their Rangers to help clear snow.

This package was custom designed for 4x4 trucks. It included a 2750 pound rated front axle, a heavy duty frame, heavy duty front springs with air bags, heavy duty front and rear shock absorbers, a 60 ampere rated alternator, an automatic transmission with cooler, and a 1620 pound payload package.

For those who towed a trailer on a regular basis there was the "Ranger Trailer Towing Package" which was rated to haul trailers up to 5100 pounds. This package came with the following equipment; a 2.8 Liter V-6 engine, automatic transmission, 3.73:1 rear axle ratio, P205/75R14 tires, super engine cooling package, 1760/1770 pound payload package, special wiring harness, rear step bumper, and a heavy duty turn signal flasher.

The "Ranger Recreation Package" was another extra cost option designed to appeal to those Ranger owners who liked to take their trucks camping. This package included front and rear stabilizer bars, heavy duty shocks, 1760/1770 pound payload package, heavy duty springs, P205/75R14 XL black-sidewall highway tires, selected rear axle ratios, 2.3 L OHC inline 4-cylinder engine, heavy duty battery, automatic transmission, super engine cooling package, and low mount Western style swing away exterior mirrors.

Ranger 4x4 buyers were also given the choice this year of ordering their trucks with either manual or automatic front axle locking hubs.

1984 Ranger Exterior Color Choices

Code	Description
1C	Raven Black
2G	Bright Bittersweet
3F	Light Blue
3L	Midnight Blue Metallic
3P	Medium Blue Metallic
51	Dark Canyon Red
53	Medium Desert Tan
8Q	Light Desert Tan
9D	Polar White
9V	Light Charcoal Metallic
9Y	Walnut Metallic
9C	Bright Copper Glow (Optional at extra cost)

Interior Color Choices
Dark Blue
Canyon Red
Tan

1985 Ranger Exterior and Interior Color Combinations

Exterior Color and Code	*Interior Color Choices*
Raven Black 1C	Canyon Red, Tan
Wimbledon White 9A	Canyon Red, Tan, Regatta Blue
Silver Metallic 1E	Canyon Red, Regatta Blue
Dark Charcoal Metallic 9W	Canyon Red, Tan
Light Canyon Red 2E	Canyon Red, Tan
Dark Canyon Red 51	Canyon Red, Tan
Light Regatta Blue 34	Regatta Blue
Bright Regatta Blue 7H	Regatta Blue
Light Desert Tan 8Q	Canyon Red, Tan
Walnut Metallic 9Y	Tan
Midnight Blue Metallic 3L	Regatta Blue, Tan
Dark Spruce Metallic	Tan

1985 was the year of the "new" as far as Ford Rangers were concerned. The following list contains all the new items that Ford blessed the Ranger with this year;

    New 2.3 Liter (140 CID) Fuel Injected Inline 4-Cylinder Engine
    New 2.3 Liter (140 CID) Turbo Diesel Inline 4- Cylinder Engine
    New Hydraulic Rear Engine Mounts on the 2.3 Liter 4-Cylinder Engine
    New Hydraulic Front Engine Mounts used on 2.3 Liter Engines with Automatic 4 Speed Overdrive Transmission
    New All Season Tread Design Tires
    New Bright Grille Surround on XLS Models
    New 4-Speed Automatic Overdrive Transmission (Ford AOD)
    New Standard 5-Speed Manual Overdrive Transmission
    New Right Hand Passenger Assist Bar
    New Improved Fuse Panel
    New "A Frame" Designed Steering Wheel
    New Fold Away Paddle Style Exterior Mirrors
    New Bright Aluminum Snap On Hub Cap
    Six New Exterior Color Choices
    New Chrome Plated Rear Step Bumper
    New Electronic AM/FM Stereo Cassette Radio
    New Power Convenience Group (Power Windows and Power Locks)
    New Horn Blow Function Moved From Stalk Mounting To A Horn Button in the Center of the Steering Wheel
    New Premium Sound Package
    New Cloth Seat Trims on Bucket Seats
    New Speed Control System
    New Engine Compartment Light
    New Emissions System Maintenance Warning Light

This is a copy of the 1984 Bronco II introductory showroom brochure.

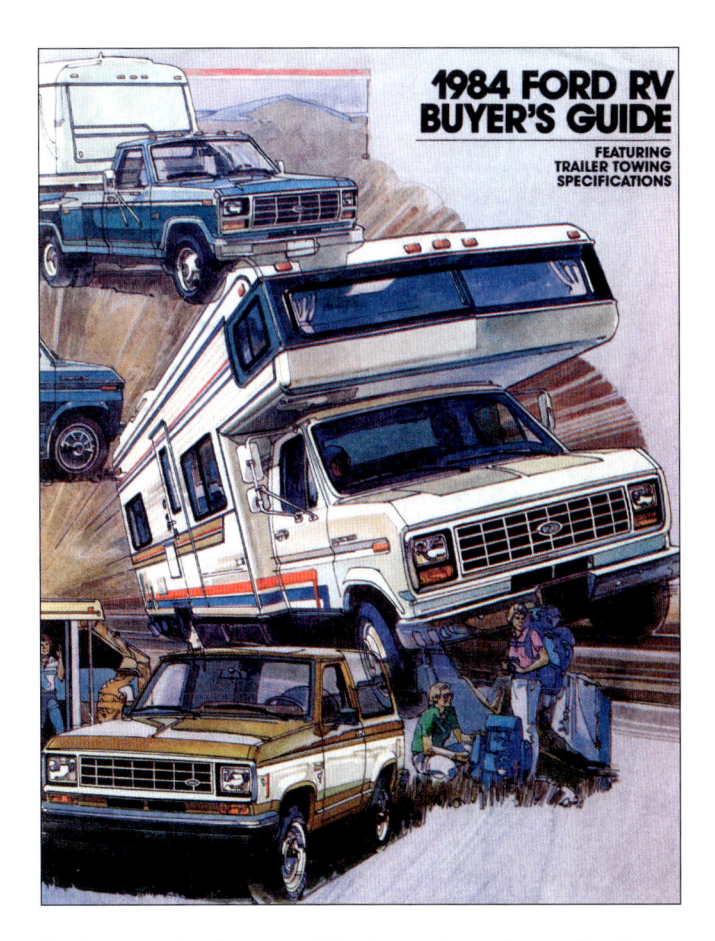

Ford was pushing their new 1984 Bronco II as evidenced by the fact that they put this new sport utility vehicle on the cover of their 1984 Ford RV Catalog.

Another magazine ad calling the Ranger "America's Truck Built Ford Tough".

Once again the Ford Ranger was available in four model versions. The standard Ranger, or base model, Ford promoted as "The Pace Setter" this year. Next on the list was the Ranger XL, which Ford referred to as their "Style and Convenience Leader".

Version number three was the Ranger XLT, which Ford called "The Ultimate in Luxury in a compact sized pickup truck. And last but not least was the Ford Ranger XLS, which Ford called "The Sporty Ranger".

Base prices for the Ranger 4x2 models ranged from $6000-$6900 while the base prices for Ranger 4x4 models ranged from $8300 to $8500.

Donald E. Petersen, Ford Motor Company President said..."I think the 1985 Ranger is an excellent example of the quality I am talking about. It combines a thrifty compact pickup size and surprising interior room with driver-oriented design and advanced engineering features. These plus built Ford tough trucks in 4x2 and 4x4 models offer you state-of-the-art pickup technology".

The standard engine for the Ranger this year was the 2.3 Liter multi-port electronic fuel injected inline 4-cylinder overhead cam engine. But for those who wanted the 2.0 Liter inline 4-cylinder engine it was still available by special order. Both engines came standard with 5-speed manual overdrive transmissions.

As we said the standard transmission for the Ranger this year was a 5-speed manual overdrive unit but for those people who didn't want to use a clutch or manually shift their gears Ford for the first time offered their excellent 4-speed automatic overdrive transmission (AOD) as an extra cost option for their Rangers.

This transmission offered the buyer 4 forward speeds from which to choose with the 4th gear being a direct mechanical hook up eliminating hydraulic slippage by about twenty five percent helping to increase mpg numbers.

Another Ford Ranger magazine ad that looks familiar. Only the punch line has been changed.

For 4x4 Ranger buyers the Ranger came equipped with a chain driven, 2-speed, part time transfer case. With the front hubs in the locked position one could "Shift-on-the-fly" pretty easily going from 2-wheel drive to 4-wheel drive and vice versa to deal with changing traction conditions. Another feature of this transfer case was that it included a constant displacement hydraulic pump for lubrication so a 4x4 Ranger could be towed at speeds up to 55 mph for any distances without having to disconnect the drive shafts or lifting the front wheels off the ground.

Earlier we listed the four different Ranger models for 1985 and all four were pretty special vehicles in their own right but for those people who wanted to make their Rangers even more special Ford's extra cost option list included 140 different items including such extras as air conditioning, bucket seats, auxiliary fuel tanks, power windows, limited slip axles, speed control, power steering, tilt steering wheels, automatic locking front hubs, and three two-tone exterior paint schemes.

Additional packages included a camper package, a handling package, snowplow package, and a trailer-towing package.

Once again Ranger buyers could choose their Rangers in a 4x2 or 4x4 layout, with long or short pickup boxes, and also they could still order a "Chassis and Cab" Ranger without a pickup box.

Looking at the specific models in detail, the base model Ranger pickup featured a black-colored front bumper, black-colored fold away exterior mirrors, chrome plated grille and bright metal windshield trim, and argent-colored wheels with bright hub caps.

Other standard equipment included a vinyl covered bench seat, an "A Frame" soft feel steering wheel, color-keyed interior trim components, stalk mounted controls, an instrument panel storage bin, glove box, an AM radio, dome light, inside hood release, a right hand assist handle, and a Day/Night rear view mirror.

Ranger XL models came with all or most of the

standard items found on the base model Rangers, plus bright metal wheel lip moldings, bright metal rear window trim insert in a molding, deluxe wheel trim, chrome plated front bumper, wood tone accent around the instrument cluster, color-keyed cloth headliner, contoured knitted vinyl bench seat cover, color-keyed seat belts, color-keyed floor mat, and passenger door courtesy lamp switch.

The top-of-the line Ranger for 1985 was still the Ranger XLT and this version had lots of upgraded deluxe trim items. Those upgraded items included a chrome plated front bumper with black-colored end caps; full-length black lower body side moldings with bright accent trim; dual accent body side paint striping, and a brushed aluminum tailgate panel.

Other XLT standard equipment included color-keyed cloth door trim panels, a contoured bench seat with a color-keyed cloth cover, color-keyed carpeting, leather wrapped steering wheel, tinted side windows, and vent windows.

Which brings us to the last Ranger model for 1985, the Ranger XLS. These XLS Rangers came with special black-colored exterior trim pieces, distinctive "XLS" tri-colored tape stripe decals, reclining cloth-covered bucket seats, a leather wrapped "A Frame" styled steering wheel, and a Gauge Package which included an ammeter, temperature and oil pressure gauges, and a trip odometer.

1985 was a pretty good year for Ford Ranger production as over 247,000 of them left Ford factories.

If you wanted a diesel engine in your Ford built truck in 1984 this magazine ad shows you could get one in a variety of shapes and sizes.

No need to explain the message of this magazine ad; the words speak for themselves.

Here is a magazine ad that promotes the toughness and quality found in the 1984 Ranger truck lineup.

This magazine advertisement talks about Ford's 4x4 vehicles for 1984.

One of the ways that Ford proved they built tough trucks was to race them, as we see in this ad showing a modified Ranger racing pickup.

Jeff Huber drove this Chief Auto Parts sponsored Ford Ranger Off-Road racing truck in the Mickey Thompson Off-Road Stadium Series in 1984.

Another Ford promotional postcard showing a Ranger XLT 4x4 truck pulling a travel trailer.

This Ranger postcard released in 1984 shows a Ranger XLT 4x2 pickup dressed up in a two-tone paint scheme and lots of deluxe trim.

27

4x4xFord!

AMERICA'S TRUCK BUILT FORD TOUGH

A truly tough pair of Ford trucks is featured in this magazine ad with a Ford F Series 4x4 beside a Ranger 4x4.

Ford built some nice looking pickup trucks back in the mid-1980s as this F-150 and Ranger 4x4 show.

Nobody 4x4's like Ford!

Tough Fords are America's best-selling 4x4's.* And for '85, high-output engines help make an even more powerful statement!

What makes Ford's best-selling 4x4's king of the off-road? **Performance.** Full-size or small-size, these tough 4-wheelers are built to handle rugged terrain...with power and toughness.

Power to spare.
Ford gives you the power to rule the off-road. With Ford's full-size F-Series pickup, the standard engine is a powerful 4.9L Six. Or you can choose from several V-8's, each one the most powerful in its class.

Like the 5.8L V-8, a high output engine with 4-barrel carb.† And the small-size Ranger's husky 2.8L V-6 option has horsepower that no other small V-6 pickup beats!

But pure power is only part of the story. Ford puts technology on your side, too! Ford's big pickup gives you the option of 5.0L V-8 with new electronic fuel-injection. And Ranger comes with a 2.3L fuel-injected 4-cylinder engine, plus 5-speed manual overdrive, standard!

Tough as they come.
Tough 4-wheeling calls for Ford's rugged Twin-Traction-Beam front suspension. The independent wheel action soaks up bumps and helps provide sure-footed traction. And now the Mono-Beam front suspension is available for those extra heavy duty jobs.

And no matter which tough Ford 4x4 you choose—the big F-Series or the small-size Ranger—you get Ford's proven 4-wheel drive system with manual or optional automatic locking hubs. Add in big Ford payload capacity and the strength of double wall box construction and you'll know why nobody 4x4's like Ford!

Best-Built American Trucks.
At Ford, Quality is Job 1. A 1984 survey established that Ford makes the best-built American trucks. This is based on an average of problems reported by owners in the prior six months on 1981-1983 models designed and built in the U.S.

Lifetime Service Guarantee.
As part of Ford Motor Company's commitment to your total satisfaction, participating Ford Dealers stand behind their work, in writing, with a Lifetime Service Guarantee. No other car companies' dealers, foreign or domestic, offer this kind of security. Nobody. See your participating Ford Dealer for details.

Dealer-installed Ranger light bar not for occupant safety.
*Based on new truck calendar year registrations thru October, 1984.
† Optional; not available in California or with manual transmission.

"My Ford Pickup & Me."   "My Ford Ranger & Me."

AMERICA'S TRUCK BUILT FORD TOUGH

"Nobody 4x4's like Ford" is a rather clever way to promote "America's Truck Built Ford Tough" in 1985.

Another magazine ad showing how Ford built tough trucks back in the mid-1980s. Here we see Manny Esquerra posing proudly beside his Ranger racing truck.

Tough guys finish first.

Ford's toughest competition Ranger has raced ahead of other small pickups to win big off-road! Ranger's got the power to make you a winner, too, with V-6 power no other small V-6 pickup beats!

Today's tough Ford Ranger 4x4 is the little truck that wins the big ones—including last year's Parker 400, one of SCORE's toughest tests!

This racing Ranger's specially modified for off-road competition, naturally. But every Ranger going, race winner or street stock, offers you tough truck features like a 2.8L V-6 with power unbeaten by any small V-6 pickup!

New 2.3L with EFI.
For '85 a new electronically fuel-injected 2.3L four has been added to the powerful Ranger lineup.

Like all Ranger 4x4's, it comes with Ford's race-tested 5-speed transmission, standard!

Above all, it's tough!
For bashing around the boonies, nobody's topped tough Ford Ranger's exclusive Twin-Traction-Beam front suspension...its proven four-wheel-drive system (choice of manual or optional automatic locking hubs). Streetwise, nobody beats the comfort of Ranger's wide cab (widest of any small pickup)...or Ranger's optional payload, 1,625 lb...or Ford's tough double-wall box construction.

Best-Built American Trucks.
At Ford, Quality is Job 1. A 1984 survey established that Ford makes the best-built American trucks. This is based on an average of problems reported by owners in the prior six months on 1981-1983 models designed and built in the U.S.

Lifetime Service Guarantee.
Participating Ford Dealers stand behind their work, in writing, with a free Lifetime Service Guarantee for as long as you own your Ford car or light truck. Ask to see this guarantee when you visit your participating Ford Dealer.

"My Ford Ranger & Me"

FORD RANGER
AMERICA'S TRUCK BUILT FORD TOUGH

This Ford Motor Company promotional photograph shows a two-tone colored Ranger XLT 4x4 posed in an off-road scene.

Here we see a white colored Ranger pickup that has been equipped with a white-colored pickup bed cap.

A set of modern 5-spoke wheels adorn this early model Ranger XLT long-bed pickup truck.

Chapter 4: 1986-1988 More Ranger Models and Introducing the SuperCab

For the 1986 model year the Ford Motor Company introduced an expanded line of Ranger models and for those who needed a longer and roomier cab Ford released their new SuperCab versions. The expanded Ranger trim levels this year included the Ranger "S", Ranger, Ranger XL, Ranger XLT, and Ranger STX.

The Ranger "S" was now Ford's base Ranger model and it was only available in a Regular Cab version. It came standard with a 2.0 Liter inline 4-cylinder engine with a maximum rating of 73 horsepower at 4000 rpm. California bound Ranger "S" models and Ranger "S" models destined for high altitude areas came standard with Ford's 2.3 Literelectronic fuel injected inline 4-cylinder engine that carried a maximum horsepower rating of 90 at 4000 rpm.

Other standard equipment found on the Ranger "S" was manual steering and brakes; and all-season, steel-belted radial tires in a P185/75R14SL size. This Ranger "S" model was designed and equipped to appeal to buyers who just wanted a bare-bones truck, so it was no surprise that the Ranger "S" model this year carried the lowest base price of any 1986 Ford Ranger.

Moving up in the Ranger lineup for 1986 was the Ranger. This upgrade came with equipment like a 2.3 Liter OHV 90 horsepower electronic fuel injected engine, power brakes on 4x2 models, P195 steel-belted radial tires, ammeter, temperature, and oil pressure gauges, and a handy trip odometer.

Other deluxe level standard equipment on the Ranger model included interval wipers, lockable glove box and storage bin, inside hood release, color-keyed seat belts, inside Day/Night mirror, halogen headlamps, bright grille and windshield moldings, black-colored fold-away exterior mirrors, and all vinyl interior trimmings.

Standard Ranger buyers had their choice of either a short Styleside pickup box (6 Foot) or an optional long Styleside pickup box (7 Foot) in either a 4x2 or 4x4 chassis arrangement. And this Ranger model was also available in either a Regular Cab or SuperCab version or as a 4x2 "Chassis Cab" model in a Regular Cab form.

Next up on the Ranger list of trim level models was the Ranger XL which this year was only available as a Regular Cab version. This Ranger came equipped with lots of upgraded items that a lot of drivers and owners wanted. Equipment like a new 2.9 Liter (179 cubic inch) multi-port electronic fuel injected V-6 engine that Ford rated at 140 horsepower at 4000 rpm. An engine that was bolted to a 5-speed manual overdrive transmission at no extra cost.

The Ranger XLs also came with upgraded deluxe

This magazine ad announces the fact that Ford's 4x4 Ranger SuperCab was voted "4x4 Of The Year" by 4-Wheel & Off-Road Magazine.

1986 RANGER MARKETING PROGRAM
RELOADING FOR LEADERSHIP!

FOR 1986, RANGER "ZEROES-IN" ON THE NO. 1 IMPORTED PICKUP FROM TOYOTA TO GIVE YOU THE COMPETITIVE EDGE!

Ford salesmen were given this promotional booklet about the same time the 1986 Rangers were introduced to teach them how to promote, and compare these trucks to the small pickups being offered by Toyota.

level interior trimmings. The contoured front bench seat was covered in a knitted vinyl material and matched with color-keyed floor mat and under a color-keyed cloth headliner.

Other Ranger XL standard equipment included interval wipers, power assisted steering, a bright metal front bumper with a protective black-colored rub strip, bright metal wheel lip moldings, windshield and rear window moldings of bright metal, and bright metal wheel trim rings.

For those Ranger buyers who wanted a top-of-the-line, luxury-trimmed pickup truck in 1986, there was the Ranger XLT SuperCab model. This super deluxe-trimmed Ranger came with such niceties as Ford's new 2.9 Liter (179 cubic inch) electronic multi-port fuel injected V-6 engine, deluxe-grade cloth seat covers, deluxe-grade cloth-covered door panels, with map pockets and carpeted lower sections, tinted pivoting rear quarter windows, 16 ounce color-keyed carpeting, a deluxe color-keyed leather wrapped steering wheel, power steering, and a two-tone exterior paint treatment.

Other standard equipment found on the Ranger XLT included an AM/FM stereo radio with 4 speakers, an under-hood light, chrome plated front bumper, interior cargo cover, dual horns, black-colored rear step bumper, low mount bright metal finished swing-away exterior mirrors, bright metal wheel trim rings, and walnut wood grain instrument panel accent trim.

All that deluxe equipment in a compact pickup truck for a base price of $8353.00 in 1986 was quite a deal. The Ranger XLT was probably the poshest mini-pickup that you could buy in that model year.

1987 FORD LIGHT TRUCK EXTERIOR COLOR SELECTIONS

Availability Legend:
R-Ranger, F-F Series, B-II-Bronco II, B-Bronco, A-Aerostar, E/CW-Econoline/Club Wagon

Code	Color	Availability
1C	Raven Black	R, F, B-II, B, A
1D	Dark Grey Metallic	F, B, E/CW
1G	Medium Silver Metallic	F, B, E/CW
14	Silver Clearcoat Metallic	R, B-II, A
2C	Midnight Canyon Red Metallic	E/CW
2E	Bright Canyon Red	R, F, B-II, B
25	Dark Cabernet Clearcoat Metallic	A
27	Medium Red	A
3J	Light Regatta Blue Metallic	E/CW
34	Light Regatta Blue	F, B, E/CW
4U	Alpine Green Clearcoat Metallic	R, B-II*
42	Alpine Green Metallic	F, B*
5U	Dark Walnut Metallic	E/CW
5Z	Light Chestnut Clearcoat Metallic	R, B-II, A
51	Dark Canyon Red	R, F, B-II, B, E/CW
7A	Spinnaker Blue	A
7B	Dark Shadow Blue Metallic	F, B, E/CW
7J	Bright Regatta Blue Clearcoat Metallic	R, B-II, A
7K	Deep Shadow Blue Clearcoat Metallic	R, B-II, A
9M	Colonial White	R, F, B-II, B, A, E/CW
9N	Desert Tan Metallic	F, B, E/CW
9R	Shadow Grey Clearcoat Metallic	R, B-II, A
9T	Light Chestnut	R, F, B-II, B, A, E/CW
9U	Dark Walnut	R, B-II, A

* Available with Eddie Bauer package only

This is a Ford color chip page showing all the colors available on Ford trucks for 1987. All the colors with an "R" code were available for the Ranger that year.

Late in the 1985 model year Ford Marketing decided to test market a special model Ranger pickup truck in the Western United States. They called this model the Ranger STX and it was equipped to go up against the sporty Toyota SR5 pickup. Because this model was so successful in this test, for 1986 the Ford Ranger STX model became available to all Ford dealers on a nationwide basis. And Ford called this new Ranger model ..."their sporty, fun to drive Ranger". The Ranger STX was available in 4x2 or 4x4 versions, in Regular Cab or SuperCab versions so it could appeal to a wide variety of customers.

This sporty-themed compact truck came loaded with a lot of standard equipment like Ford's 2.9 Liter electronically controlled multi-port fuel injected V-6 engine, power assisted steering, unique accent "STX" themed exterior two-tone paint scheme, black-colored foldaway exterior mirrors, black painted grille with bright metal surround trim, black-colored bumpers, rear cargo light, and cloth-covered reclining bucket seats.

Ranger SuperCab STX models came equipped with Captains Chairs with power lumbar supports, and an AM/FM stereo radio system.

Regular Cab 4x2 STX models came standard with a Handling Package suspension system, gas pressurized shock absorbers, front stabilizer bar, and P205/70R14SL Eagle GT raised-white-letter tires. While the Ranger STX 4x4 trucks came standard with skid plates and P215/75R15SL, raised-white-letter, off-road spec tires.

1986 Ranger Exterior Color Choices

Dark Canyon Red
Medium Silver Metallic
Raven Black
Colonial White
Desert Tan
Light Regatta Blue
Dark Gray Metallic
Dark Spruce Metallic
Silver Clearcoat Metallic
Dark Canyon Red Metallic
Dark Shadow Blue Metallic
Desert Tan Metallic
Bright Regatta Blue Clearcoat Metallic
Light Chestnut Clearcoat Metallic
Dark Walnut Metallic
Bright Canyon Red

SuperCab Exterior Colors

Silver Clearcoat Metallic
Medium Canyon Red Clearcoat Metallic
Light Chestnut Clearcoat Metallic
Bright Regatta Blue Clearcoat Metallic

Interior Colors

Regatta Blue
Canyon Red
Chestnut

Regular Cab Rangers came in two wheelbase lengths, the shorter version had a wheelbase of 107.9 inches and a front bumper to back-of-the-cab length of 99.3 inches. The longer version sat on a wheelbase of 113.9 inches and the same front bumper to back- of-the-cab length of 99.3 inches. Overall lengths of the short and long wheelbase Regular Cab models sat at 175.6 inches and 187.6 inches respectively.

The SuperCab models on the other hand sat on

a 125 inch wheelbase and a front bumper to back-of-cab length of 116.4 inches. SuperCabs were only available with a 6-foot pickup bed (actually the pickup bed length was 76.3 inches) for an overall length of 192.7 inches.

1986 Ranger Base Curb Weights

Ranger Regular Cab 4x2 (107.9 inch wheelbase)	2600 pounds
Ranger Regular Cab 4x2 (113.9 inch wheelbase)	2638 pounds
Ranger SuperCab 4x2 (125 inch wheelbase)	2842 pounds
Ranger Regular Cab 4x4 (107.9 inch wheelbase)	2833 pounds
Ranger Regular Cab 4x4 (113.9 inch wheelbase)	2889 pounds
Ranger SuperCab 4x4 (125 inch wheelbase)	3065 pounds

For the 1983 and 1984 model years the Ford Ranger was the top-selling vehicle in the compact truck market segment; but in 1985 Toyota edged out the Ranger to be the top selling compact pickup truck in the United States. So for the 1986 model year Ford Marketing decided to concentrate on competing against the entire Toyota compact truck lineup.

For the sporty truck buyer there were the Ranger STX models that went up against the Toyota SR5 Sport Truck in both 4x2 and 4x4 versions. Both manufacturers offered these models in Regular Cab and SuperCab or extended cab versions.

Luxury and personal transportation buyers had the Ranger XLT and XL versions to choose from while Toyota didn't offer their buyers any direct comparative models. Businessmen and buyers who needed work trucks could choose between the Ranger and Ranger "S" models at a lower price than the higher tiered trim levels while Toyota offered their own work type trucks in either a standard or deluxe level trim.

Ford also offered an overhead valve 4-cylinder turbocharged diesel engine this year that displaced 2.3 Liters or 143 cubic inches. This engine featured electronic fuel injection, a compression ratio of 21.1:1, a net horsepower rating of 86 and a torque rating of 134 lb/ft at a low 2000 rpm.

The standard transmission for all Ford Rangers this

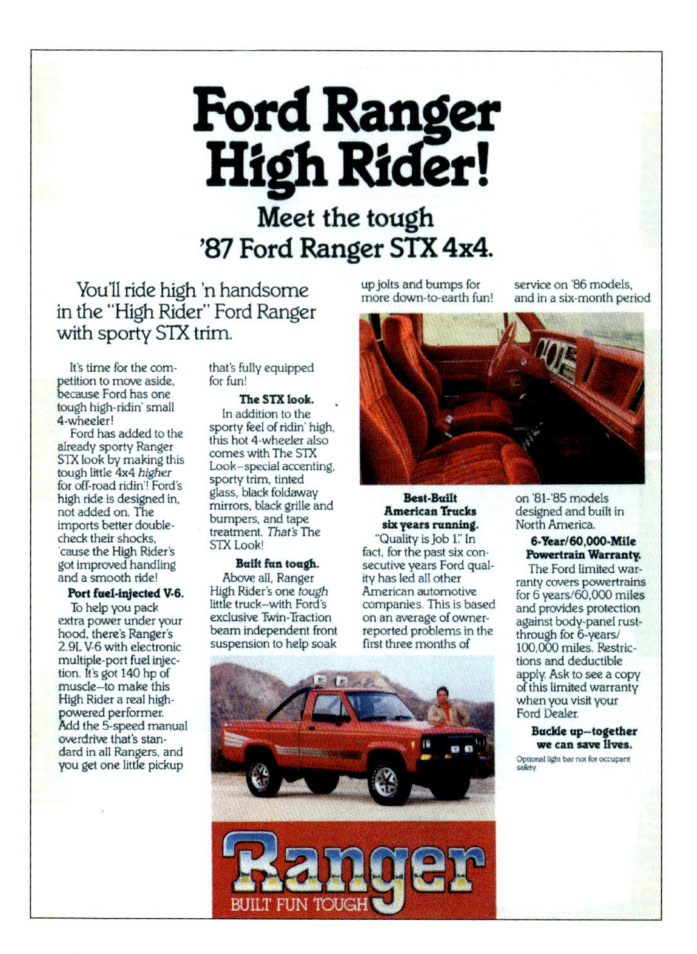

For those truck buyers looking for a "tough" looking truck, Ford offered them the chance to buy their new 1987 Ranger STX 4x4 "High Rider" truck model.

year was a 5-speed manual with overdrive that was supplied to Ford by either Mitsubishi or Mazda. The only optional transmission offered for the Ranger was Ford's A4LD 4 speed automatic overdrive unit that was priced at $793.00.

Four-wheel drive Rangers came standard with manual locking front hubs. For those who wanted an automatic locking front hub system, Ford offered their "Electric Shift Touch Drive Option" which featured control buttons mounted in an overhead console unit. For base models like the Ranger and Ranger "S", this option was priced at $168.00 while the price for this option on the higher tiered models like the STX, XLT, and XL the price was set at $101.00.

Once again Ford offered a "Camper Package" for the Ranger Regular Cab version in both 4x2 and 4x4 chassis layouts. This optional package was priced at $110.00 and it included front and rear stabilizer bars, heavy-duty gas pressurized shocks, heavy-

One of the magazine ads that Ford used to sell the Ranger in 1987 featured this Ranger SuperCab in a scene with some hot air balloons.

FORD RANGER...FUN WITH ALL THE TRIMMINGS.

FORD RANGER
BUILT FUN TOUGH

duty front springs, a 1760 pound payload package, P205/75R14SL black-wall, all-season radial tires, and a choice of axle ratios.

In order to save their customers some money and offer them a variety of equipment to tailor make their Rangers Ford offered eight "Preferred Equipment Packages" for their Ranger trucks. By combining different equipment into these packages a customer was able to save some money over what it would cost for each individual item if ordered separately. In doing so company was able to give the customer a better equipped vehicle at a cheaper price making it more appealing to the buyer.

With their new emphasis on providing their customers with an expanded range of "Preferred Equipment Packages" the company decided to whittle down some of their optional equipment items that had been available to buyers of earlier Rangers. Some of the items that were deleted from the optional equipment list were the "Trailer Towing Package", "Snowplow Special Package", "Regular and Special Two-Tone Paint Packages", tie-down hooks, floor consoles, "Light Group", "Exterior Protection Package", white painted sport wheels, tow hooks, and the "XLS Trim Package".

During the 1986 model year 4 Wheel and Off Road Magazine picked the Ranger Super Cab as their "Truck of the Year", which was quite an honor for such a new model.

Targeting Toyota proved to be quite successful for Ford Marketing in 1986 because at the end of the model year Ford's Ranger was once again the top-selling compact pickup truck in the United States.

Since 1986 was a pretty good year for the Ranger, Ford decided to make the Ranger even more appealing to compact truck buyers to help their truck stay at the top of the sales charts. The company also decided to cut back on Ranger model choices by elimination the Ranger XL models from the lineup. They also renamed the Ranger model by now calling it the Ranger Custom. The Ranger "S" was still the cheapest model Ranger in the mix and the Ranger XLT was still the Ranger to buy for those buyers willing to pay a little extra for a compact truck with a taste of luxury. And for those customers who liked their Rangers to be on the sporty side there were still the Ranger STX models from which to choose.

The Ranger "S" was still the mini-truck choice for buyers looking for a basic good-looking truck that was reliable and dependable for the least amount of money.

These trucks came standard with a 2.0 Liter inline

This Ranger SuperCab pickup features a nice two-tone paint job and a fancy looking pickup truck bed cap.

4-cylinder engine, manual brakes, and P185/75R14 SL all-season, steel-belted radial tires. For Rangers sold in California or high altitude areas, the base engine found in these vehicles was the 2.3 Liter inline 4-cylinder engine.

For a little more money the Ranger Custom was a better choice for most Ranger buyers. That extra cost bought you a truck with power brakes, gas pressurized shocks for a better ride, halogen headlights, interval windshield wipers, instrument panel gauge cluster, color-keyed seat belts, and a vinyl covered bench seat. And unlike the Ranger "S" the Ranger Custom was available with either a Regular Cab or a Super Cab version.

Besides offering a bigger and roomier cab the SuperCab model came standard with power steering, tinted glass, and a cloth-covered headliner.

Since there was not a Ranger XL model this year, the next step-up model in the Ranger lineup was the Ranger XLT; a truck that was designed to appeal to owners who wanted a classier looking pickup that was full of extra deluxe items.

In the Ranger XLT Regular Cab version, you would find a contoured bench seat with a knitted vinyl cover, plush looking color-keyed carpeting, a color-keyed cloth headliner, wood tone instrument panel

Another Ranger XLT pickup painted in a two-tone combination.

accents, an AM/FM stereo radio with digital clock, Deluxe Two-Tone exterior paint, low mount swing away bright metal exterior mirrors, a chrome plated front bumper with rub strip, a chrome plated rear step bumper, deluxe wheel trim pieces, bright metal wheel lip moldings, and a bright metal tailgate panel.

The Ranger XLT SuperCab came standard with most of the items found in the Regular Cab model plus a 60/40 split bench seat in front and dual vinyl-covered rear bench seats.

The Ranger STX was once again the sportiest looking Ranger of the mix in 1987. If you were

This Ranger XLT Regular Cab pickup truck found in Colorado features a two-tone deluxe paint job and a set of upper body bed rails.

looking for a great looking, sporty fun to drive, Ranger the STX model was definitely the one to have. Available in either 4x2 or 4x4 layout, Regular Cab or SuperCab version, this truck was designed to appeal to younger Ranger buyers. This year Ford referred to their Ranger STX models as their top-of-the-line vehicles and it was easy to see why. The STX models came with reclining cloth-covered bucket seats in the Regular Cab version and with dual cloth-covered Captains Chairs with power lumbar supports in the SuperCab version.

Other STX standard equipment included cloth-covered door panels with map pockets, a leather covered steering wheel, an AM/FM stereo radio with digital clock and 4 speakers, pivoting front vent

Racing Ranger team pickups used this type of tri-color paint schemes in the late 1980's. This street driven Ranger pickup is set up like an off-road racer of that period with its bed mounted roll bar, spare tire, net, and tubular rear bumper.

Base Prices of 1987 Rangers

Ranger "S" Styleside with Short Pickup Bed	$6788.00
Custom Ranger (2.3 L 4-cylinder) Styleside with Short Bed	$8079.00
Custom Ranger (2.3 L 4-cylinder) Styleside with Long Bed	$8240.00
Custom Ranger (2.3 L 4-cylinder) Styleside with Short Bed and 4x4	$10,519.00
Custom Ranger (2.3 L 4-cylinder) Styleside with Long Bed and 4x4	$10,682.00
Custom Ranger SuperCab Styleside with Short Bed4x2	$9241.00
Custom Ranger SuperCab Styleside with Short Bed4x4	$11,757.00
Ranger XLT (2.3 L 4-cylinder) Styleside with Short Bed Preferred Package	$8561.00
Ranger XLT (2.3 L 4-cylinder) Stylesidewith Long Bed Preferred Package	$8722.00
Ranger XLT (2.3 L 4-cylinder) Styleside with Short Bed and 4x4	$10,831.00
Ranger XLT (2.3L 4-cylinder) Styleside with Long Bed and 4x4	$10,994.00
Ranger XLT (2.3 L 4-cylinder) SuperCab Styleside Pickup	$9723.00
Ranger XLT (2.3 L 4-cylinder) SuperCab Styleside Pickup and 4x4	$12,069.00
Ranger XLT (2.9 L V-6 Engine) Styleside Pickup with Short Bed	$9758.00
Ranger XLT (2.9 L V-6 Engine) Styleside Pickup with Long Bed	$9837.00
Ranger XLT (2.9 L V-6 Engine) Styleside Pickup with Short Bed and 4x4	$12,028.00
Ranger XLT (2.9 L V-6 Engine) Styleside Pickup with Long Bed and 4x4	$12,191.00
Ranger XLT (2.9 L V-6 Engine) SuperCab Styleside Pickup with Short Bed	$10,920.00
Ranger XLT (2.9 L V-6 Engine) SuperCab Styleside Pickup with 4x4	$13,266.00
Ranger STX Styleside Pickup with Short Bed and Preferred Package 865B	$9851.00
Ranger STX Styleside Pickup with Long Bed and Preferred Package 865B	$10,012.00
Ranger STX Styleside Pickup with Short Bed and Preferred Package 865B and 4x4	$12,454.00
Ranger STX SuperCab Styleside Pickup with Preferred Package 865A	$11,013.00
Ranger STX SuperCab Styleside Pickup with Preferred Package 865A and 4x4	$13,271.00

Another Ranger XLT Regular Cab pickup with a bumper mounted grille bar, a two-tone paint job, and dark tinted windows.

Popular 1987 Ranger Options With Prices

Air Conditioning	$744.00
Chrome Package (STX Only)	$259.00
Super Engine Cooling Package	$55.00
Electric Shift 4x4 Touch Drive	
Custom Ranger Trim Regular Cab	$171.00
Other Trims Regular Cab	$104.00
Tinted Glass	$46.00
Camper Package (4x2 Models)	$190.00
Camper Package (4x4 Models)	$110.00
Handling Package (4x2 Except SuperCab Models)	$76.00
Handling Package (4x4 and SuperCabs)	$38.00
Two-Tone Deluxe Exterior Paint Custom Rangers	$234.00
Clearcoat Paints (Custom Models)	$80.00
Electronic AM Radio (Ranger S)	$61.00
Electronic AM/FM Stereo Radio/Clock (Custom Rangers)	$93.00
Electronic AM/FM Stereo Radio with Clock and Cassette Player (Custom Rangers)	$193.00
Electronic AM/FM Stereo Radio with Clock and Cassette Player (XLT and STX Models)	$100.00
Vinyl Covered Rear Jump Seats (SuperCab)	$226.00
Sport Appearance Black Package	$833.00
Sport Appearance Package Bright Alloy	$1091.00
Speed Control with Tilt Steering Wheel (Requires Power Steering Option)	$279.00
Power Steering (Custom and Ranger "S")	$274.00
Heavy Duty Front Suspension Package (Regular Cab 4x4 Models Only)	$151.00
Tachometer (Standard on STX Models)	$55.00
Cast Aluminum Wheels (4x4 Only)	$205.00
White Sport Steel Wheels (Custom)	$121.00
White Sport Steel Wheels (XLT and STX)	$32.00
Sliding Rear Window	$78.00

windows, a trip odometer, tachometer, and other gauges, and an electronic fuel injected 2.9 Liter V-6 engine.

On STX 4x2 models Ford equipped them with a special handling package that was built around a set of P205/70R14SL Eagle GT white-letter tires.

STX Rangers came one of two ways this year with either a unique exterior body tape treatment on 4x4 models or a lower body two-tone accent paint treatment with unique tape striping for 4x2 models.

All Ranger STX models came with black-colored foldaway mirrors, black-colored grilles with bright metal surround trim, black-colored contoured front bumpers and black-colored rear step bumpers.

Four wheel drive Ranger STX SuperCab models came with their own special handling package which included a set of P215/75R15 SL white-letter-outlined, off-road tires.

For years some Ford truck buyers liked to personalize their trucks by fitting them with lift kits and other aftermarket accessories. Ford noticed that most of their customers modifying their trucks in this fashion were young and Ford decided to come out with a special Ranger STX model to appeal to them.

Ford called this youth-oriented Regular Cab special

model their "Ranger High Rider 4x4 STX Truck".

The primary feature of these trucks that made them stand out from the rest of the Ranger pack was their 1 1/2 inch higher ride height which was accomplished by Ford engineers revising the truck's suspension geometry, adding a set of heavy duty gas shocks, and installing some softer rated springs. Adding these special features to the regular equipment Ford provided with their STX trucks gave them an even sportier looking truck to appeal to young buyers. And those buyers didn't have to modify their trucks because Ford already did it for them.

Yes, the "High Rider STX Ranger" Regular Cab truck models were very sporty looking trucks but for those buyers who wanted an even sportier looking Ranger truck Ford offered buyers their "Sport Appearance Package Option" which came with a black-colored bed mounted roll bar/light bar with off-road lights, a black-colored grille guard with mounted fog lamps, and a black-colored tubular rear bumper. If a buyer didn't like black-colored pieces he or she could choose Ford's "Bright Alloy Sport Appearance Package" which included all the items of the other package in a silver color. Dressed out either way, a Ranger STX High Rider 4x4 made a very bold statement at the local hamburger drive in restaurant or out on the trail.

Other changes seen on the 1987 Ranger models included a "low engine oil warning system", a new brushed chrome "Ranger" emblem on a black background, and more galvanized steel body panels to help reduce rusting problems. Another change involved the SuperCab Ranger models which were no longer available with a 2.3 Liter inline 4-cylinder engine and 5-speed manual transmission combination.

When the 1987 model year came to a close Ford truck executives had to be happy with the performance of the Ranger both on the sales charts as well as with their production numbers. By the time the last 1987 Ranger left the Ford factories, 318,000 of them had rolled off the line.

Once again for the 1988 model year the Ranger lineup included the Ranger "S", Ranger Custom, Ranger XLT, and the top-of-the-line Ranger STX. The base model Ranger was still the Ranger "S" which came with a 2.0 Liter, 2 barrel carbureted inline 4- cylinder engine and a 5-speed manual overdrive transmission, manual brakes, and manual steering. It was basically a no-frills vehicle that was the cheapest Ford pickup offered in 1988.

For those buyers who wanted a low cost compact truck with a few comfort and convenience items Ford offered a new upgraded pickup model called the Ranger "S Plus". Like the Ranger "S" the Ranger "S Plus" was only available as a Regular Cab model. The Ranger "S Plus" was also powered by the same 2.0 Liter, 2 Barrel carbureted inline 4-cylinder engine rated at 80 horsepower at 4200 rpm. The Ranger "S Plus" came with the same standard equipment items of the Ranger "S" plus a black painted rear step bumper and a color-keyed headliner.

Unless you really wanted Ford's lowest priced, half -ton rated pickup truck, the Ranger Custom was a better buy when you considered the content of the extra equipment offered on this model when compared to what your dollars bought with the Ranger "S" and Ranger "S Plus". The Ranger Custom was available again in either a Regular Cab or SuperCab version. Both of these trucks came standard with power brakes, black-colored rear step bumper, tinted glass, interval windshield wipers, , a full complement of instrument panel gauges, a 2.3 Liter (90 horsepower) electronic fuel injected inline 4-cylinder engine, AM/FM stereo radio, and more.

Next up in the lineup again were the Ranger XLT trucks in Regular Cab and SuperCab models. These Rangers came standard with a lot of extra value items. Items like a chrome plated grille with rub strip, a chrome plated rear step bumper, black-colored fold away mirrors, chrome plated grille and trim, tinted glass, Deluxe Two-Tone Exterior Paint scheme, black front spoiler, bright metal hub caps and wheel trim rings, a cloth-covered interior treatment, and a brushed aluminum tailgate panel.

Other items included in the XLT trim package were vinyl covered rear jump seats in the SuperCab models, vinyl color-keyed door panels, color-keyed cloth seat covers, color-keyed cloth headliner, color-keyed carpeting, wood-toned instrument panel accents, power steering (Regular Cab 4x4 and SuperCab), Light Group, rear locking storage compartment in SuperCab without rear jump seats, sliding rear window, electronic AM/FM stereo radio with clock, tachometer, heavy duty battery (SuperCab only), and the 2.3 Liter EFI inline 4-cylinder engine.

The boldest looking Ranger for 1988 had to be the

This photograph shows how a regular street Ranger pickup got turned into a modified Ranger race truck at Bill Stroppe's shop in California.

Ranger STX in Regular Cab or SuperCab versions, in 4x2 or 4x4 forms. Being loaded with lots of standard equipment made these models, Ford's top-of-the-line sporty compact trucks, appeal to a wide variety of buyers.

The STX Ranger standard equipment list included the 2.9 Liter EFI V-6 engine rated at 140 horsepower, front and rear stabilizer bars; P215 white-letter, all-season radial tires; black-colored front and rear bumpers, black-colored foldaway mirrors, tinted glass, "2WD" tape treatment, black-colored front spoiler, bright wheel trim rings and lug nuts, deep dish cast aluminum wheels, dual cloth-covered Captains Chairs in SuperCabs, and cloth-covered bucket seats in Regular Cab models.

Other standard Ranger STX equipment items included vinyl covered rear jump seats (SuperCabs); color-keyed, cloth-covered door panels with map pockets and carpet covered lower areas; color-keyed carpeting, right hand visor mounted vanity mirror, brushed pewter instrument panel accents, tachometer, deluxe leather wrapped steering wheel, power steering, cargo cover (SuperCab), and "High Rider Package" on Regular Cab 4x4 models.

In 1987 Ford built a limited number of Ranger GT models to test market in California. Though the number of these trucks was limited, Ford got a lot of positive feedback about this sporty looking special model; and so for the 1988 model year Ford decided to offer the "Ranger GT" on a national basis as an option for the Ranger STX 4x2 trucks.

These Ranger GTs came with all the standard STX equipment plus a unique sculpted front air dam with mounted fog lamps, unique front and rear cladding panels instead of regular bumpers, "ground effects" rocker panel covers, and wheel well spats.

The Ranger GT and the Ranger 4x4 STX "High Rider" packages were definitely geared to appeal to young sport truck enthusiasts. If the "High Rider" Rangers and the Ranger GTs didn't look sporty enough, 4x4 Ranger buyers could still add some pizzazz to their trucks by adding the "Sports Appearance Packages."

Once again, when the 1988 model year came to a close Ford's Ranger was the top-selling compact truck in the United States.

1988 Ranger Ford Dealer Accessories

Automatic Day/Night Mirror	$165.00	Bed Mounted Rally Roll Bar	$152.96
Anti-Theft System (Standard)	$49.95	Off-Road Lights (Rally Bar Mounted)	$119.23
(Deluxe)	$124.95	Grille Brush Guard	$71.09
Tot Guard Safety Seat	$60.44	Grille Guard	$94.39
Infant Safety Seat	$36.45	Tubular Rear Bumper	$139.77
Engine Block Heater	$23.15	Fender Flares	$75.44
Cellular Mobile Telephone	$1293.75	Bumper Guards	$46.29
Fire Extinguisher	$30.75	Hood Deflector Plastic Shield	$38.84
Multi-Purpose Spotlight	$27.81	Ford Command Cruise Control	$143.71
Portable Refrigerator/Warmer	$179.00	Rear Window Defroster	$49.95
Cab Storage Organizer	$97.33	Pickup Box Top Rails	$100.80
Visor Vanity Mirror	$5.39	Class II Trailer Hitches	$119.33
Front Carpeted Floor Mats	$24.21	Class III Trailer Hitch	$149.39
Body Side Protection Moldings	$44.21	Ranger Box Tire Carrier	$44.00
License Plate Frame	$5.85	Ranger Vinyl Tire Cover	$44.00
Door Edge Guards	$10.47	Rear Deck Cargo Lamp	$36.95
Saddle Blanket Seat Covers	$47.91	Spare Tire Lock	$13.11
Wheel Mud Flaps	$12.33	Wheel Lock Nuts	$25.24
Black Vinyl Heavy Duty Front End Cover	$91.16	Swing Lock Low Profile Mirror	$35.63
Fog Lamps	$73.33	Nyracord Pickup Bed Mat	$109.95
Sliding Rear Window	$103.64	Nyracord Pickup Tailgate Panel	$34.95
Aerodynamic Truck Cab Wing	$139.95	High Density Polyethylene Bed Liner	$226.67
Tonneau Cover Vinyl Coated Polyester	$151.88	Cargo Bed Stabilizer Kit	$59.95
Rear Step Bumper	$116.75	Aluminum Tailgate Top Protector	$17.83
Running Boards	$120.60	Bed Wheel Lip Tool Box	$79.95
Exterior Side Stripes	$32.00	Crossover Bed Tool Box	$138.32
Hood Tape Stripes	$33.44	Rear Window Bed Grille Guard	$120.71

"The Street Machine that Isn't Confined to the Street" was a good way to promote the Ranger STX 4x4 pickup truck in 1989.

Chapter 5: A New Look To America's Best Selling Compact Truck

Ford introduced their 1989 Ranger models as being "America's Best Selling Compact Truck That Is Fun Tough". Fun and tough were two great words to use when describing these trucks. They were fun to drive while at the same time they were some of the toughest little trucks on the market. You name the job… and chances were good that a Ranger pickup could handle it.

Being fun and tough were not the most notable features of these new trucks. The most notable feature of these new 1989 Rangers was their all-new look. In fact, these new 1989 Rangers featured the first major redesign of the marque since the Ranger was introduced back in early 1982.

The first thing you noticed about these new Rangers was their more aerodynamic front end treatment; a sleeker front end that included a set of aerodynamic halogen headlamps with integral parking and turn signal lights, a wrap around front bumper and grille, and a set of modified front fenders that integrated all these separate elements into a very pleasing, more modern look.

Then there was the restyled sloping hood line and a hood that featured some wind splits to help air flow more freely up and over the hood. Another new front end feature was the two cooling slots that were located in the center of the upper front bumper area.

In addition to changing the exterior looks of these compact trucks Ford designers also worked magic on the Ranger interiors to make them look more appealing and more comfortable.

They spent a lot of time reworking the instrument

panel to make it more attractive and more ergonomically functional. They redesigned the analog gauges to make them easier to read and they also relocated all the controls to be an easy reach for the driver. Those designers also moved the turn signal switch control, windshield wipers and washer, headlamp high beams, and the "flash to pass" feature were all combined on a column mounted lever to the left of the driver. To the driver's right, in separate stacked pods, were the "Touch Drive" buttons and lamps for the electric shift transfer case 4x4 controls, the radios, and the heater and defroster controls. Another new feature found on these redesigned instrument panels was some demisters that helped to keep the side windows from fogging up.

All Ranger models this year featured a set of interval wipers as standard equipment and for Rangers equipped with automatic transmissions, there was a redesigned automatic shift control lever mounted on the steering column.

Ford also introduced a redesigned and re-engineered inline 4-cylinder overhead valve EFI engine that became the standard engine for all Rangers except the STX models. This redesigned 2.3 Liter inline 4-cylinder engine featured a new twin spark cylinder head, new computer designed intake ports which increased the amount of air flowing through these passages thus increasing the efficiency of this new engine.

To add to the increased efficiency of this new head design, Ford engineers also redesigned the intake manifold which provided a more balanced flow of air to each cylinder's combustion chamber which in turn increased the power output of this engine to a maximum rating of 100 horsepower at 4600 rpm.

These new base Ranger 4-cylinder engines also came equipped with a new "Distributor -less Ignition System (DIS). The regular style distributors found on earlier Rangers and other vehicles for decades was replaced by a "pulse wheel" that was mounted to the front of the crankshaft which in turn worked with a magnetic field sensor that sent a spark through wires up to a set of separate coils that were mounted atop each spark plug. This system replaced the regular distributor, cap, and rotor found in the conventional ignition systems previously used.

This engine replaced the original 2.3 Liter inline 4-cylinder engine used in earlier Rangers. Since this new engine was the base engine for the Ranger Ford

The Ranger GT was indeed a "Fun to Drive" and an eye-catching youth-oriented truck with its unique dress up equipment.

dropped the 2.0 Liter inline 4-cylinder carbureted engine used in the Ranger "S" and "S Plus" versions from previous model years. Ford also dropped the optional inline 4-cylinder diesel engine that had been available since the early days of the Ranger. The only two engines Ford offered in the 1989 Rangers was this new 4-cylinder engine and their 2.9 Liter V-6 that still had a maximum horsepower rating of 140 at 4600 rpm.

Another new feature found on all new Rangers this year was an improved braking system that included an anti-lock feature on the rear brakes that was designed to minimize rear wheel lockup in most situations.

Four wheel drive Rangers this year were equipped with an improved front "Twin Traction Beam Axle System" with a new front stabilizer bar that was repositioned forward of the front axle which gave these trucks better handling characteristics.

Once again the base model Ranger was the Ranger "S" which shared some of the same standard equipment found in the pricier "S Plus" and Custom models. Items like a black painted front bumper

This is one sweet looking 1989 Ranger XLT Regular Cab short-bed pickup truck.

and grille, an all-vinyl trimmed interior, new wheel designs, a 2.3 Liter/100 HP engine, 5-speed manual /overdrive transmission, an electronic AM radio, and many more items.

Stepping up and into a Ranger "S Plus" or Custom got you a black-colored rear step bumper at no extra cost and a fancier looking vinyl interior in the Custom.

For the owner who wanted a compact truck that handled like a sports car this year, the Ranger GT was now a separate model rather than just being an upgrade package for the Ranger STX. Ford referred to the Ranger GT Regular Cab model as a "unique 2-wheel drive sport pickup designed in the spirit of Ford Motorsports".

The Ranger GT came with a unique front bumper cover and spoiler combination that incorporated a set of fog lamps, a unique rear bumper cover, "ground effects" rocker panel covers with wheel spats and a 140 horsepower rated 2.9 Liter V-6 engine under its hood. Other equipment found on the Ranger GT was a standard limited slip rear performance axle, front and rear stabilizer bars, heavy duty gas pressurized shocks, and a set of P215/70R14 white-letter, steel-belted performance radial tires.

Inside the Ranger GT, the buyer had their choice between a set of reclining sport bucket seats or a set of more comfortable Captains Chairs. Also included in the Ranger GT interiors were a set of fancy door trim panels that featured cloth inserts, map pockets, and carpeted lower sections. The standard sound system in the Ranger GTs was an AM/FM stereo radio with 4 speakers. Ranger GTs also came equipped with tachometers and leather wrapped steering wheels.

For those Ranger buyers looking for a truck with an extra touch of special value the Ranger XLT was the top choice in 1989. These trucks came with all the standard equipment found on the Ranger Custom plus deluxe wheel trim, chrome plated bumpers, a chrome plated grille, and a deluxe two-tone exterior paint treatment.

Inside the 1989 Ranger XLT there was a cloth-covered, three passenger 60/40 split bench seat, an AM/FM stereo system with a cassette player and 4 speakers, color-keyed carpeting, fancy looking door panels, a tachometer, Courtesy Light Package, and a sliding rear window. The SuperCab Ranger XLTs also came with a set of rear jump seats.

The top-of-the-line Ranger pickup truck for 1989 was the Ranger STX in 4x2 or 4x4, Regular Cab and SuperCab models. The STX 4x2 models came with Ford's 2.9 Liter V-6 engine, a 5-speed manual overdrive transmission, limited slip rear axle, stabilizer bars front and back, heavy duty gas shocks, and P215/70R Eagle GT steel-belted performance radial tires.

Ford referred to their 1989 Ranger STX models their "Sport Riders" because of its unique suspension system which featured 21 unique components that helped to increase the ride height of these vehicles and improve their performance both on and off road.

The 4x4 Ranger STX models came standard this year with P215/75R15 white-outline-lettered performance tires wrapped around a set of deep-dish aluminum wheels mounted on a limited slip rear axle.

All 1989 Ranger STX models came with a set of cloth-covered reclining bucket seats with lumbar supports or a set of Captains Chairs. STX door trim panels came with cloth inserts, map pockets, and lower sections that were carpeted. Also found in the interiors of STX models were tachometers, an AM/FM stereo system with 4 speakers, carpeting on the floor, and a leather wrapped steering wheel.

Other standard STX equipment included special "STX" sport striping, a black-colored tubular rear bumper (4x2 models), a black-colored rear step bumper on 4x4 models, a black-colored grille, and a black-colored front bumper.

And if your Ranger STX model wasn't exciting enough you could always order the "Sport Appearance Package" in either black or silver alloy colors that added a front brush/grille guard with fog lamps, a tubular rear bumper, and a bed mounted Rally Bar with mounted off-road lights.

"Ford Ranger is the best selling compact pickup truck in America and it is no wonder. Ford Ranger offers a lot to today's contemporary small truck buyer". So said a Ford promotional statement when the new 1990 Rangers made their debut in October of 1989.

The 1990 Ranger lineup consisted of the Ranger "S", Ranger "S Plus", Ranger Custom, Ranger XLT, and the Ranger STX. And those Rangers were available in the following exterior colors;

Sandalwood Clearcoat Metallic
Light Sandalwood Clearcoat Metallic
Light Sandalwood
Cabernet Red
Scarlet Red
Hunter Green Clearcoat Metallic
Crystal Blue Clearcoat Metallic
Raven Black
Silver Clearcoat Metallic
Shadow Gray Clearcoat Metallic
Colonial White

Interior color choices were Crystal Blue, Scarlet Red, Medium Gray, and Light Sandalwood.

The standard engine for the Ranger "S", Ranger "S Plus", Ranger Custom in both 2- wheel drive and 4-wheel drive was the dual spark plug 2.3 Liter EFI inline 4-cylinder. While the standard engine for the Ranger XLT, Ranger STX, and the Ranger Custom 4x4 SuperCab was the 2.9 Liter EFI V-6 engine.

For those Ranger buyers who were looking for more power and more torque, Ford introduced a bigger V-6 engine option for the Ranger and Aeromax mini-van this year. This new, extra cost, engine option was built by Ford of Germany and it had a displacement of 4.0 Liters or 245 cubic inches. This new engine was based on the same engine architecture as the 2.9 Liter V-6 and it would be the largest engine ever to be offered by Ford in a Ranger.

This new 4.0 Liter V-6 engine was more than simply a bored and stroked version of the 2.9 Liter V-6 engine.

The new 4.0 Liter V-6 engine featured the following items not found on the smaller 2.9 Liter engine;

• A new EEC-IV Processor which used a Mass Air Flow Sensing System and a simultaneous double fire injector EEC-IV strategy to control engine functions.

• A larger bore and a longer stroke with a new package of rings designed to reduce tension and power-robbing friction.

• This larger engine also featured a higher deck height of 20 mm and a widened oil pan rail to accommodate longer crank throws.

• Revised intake and exhaust ports.

• Redesigned combustion chambers and new larger valves.

• New pushrods and a new camshaft design.

• New Electronic Distributor Ignition System (EDIS) with

• 3 sets of coil packs mounted on the right rocker cover.

• A new Ford permanent magnet design lighter weight starter.

• A new serpentine accessory belt system that incorporated a single belt to run all the accessories replacing the three belt system used on the other Ranger engines.

This engine option in 1990 was only available with a modified and upgraded Ford 4- speed automatic overdrive transmission. It was also offered as an option for all XLT and STX Rangers this year.

1990 V-6 Engine Specifications

2.9 Liter	4.0 Liter
Gasoline Engine	Gasoline Engine
V-6 Cylinder Layout	V-6 Cylinder Layout
Displacement: 2.9 Liter/179 Cubic Inches	4.0 Liters/245 Cubic Inches
Cylinder Head: 2 Valves	2 Valves
Fuel System: Multi Port EFI	Multi Port EFI
BorexStroke: 3.66x2.83 inches	3.95x3.32 inches
Compression Ratio: 9.0:1	9.0:1
Maximum Horsepower:140@4600 rpm	160@ 4200 rpm
Maximum Torque: 170 lb/ft @ 2600 rpm	225 lb/ft @ 2400rpm

Another Ranger XLT Regular Cab pickup truck wearing a color-keyed pickup bed cap.

At the start of the 1990 model year the Ranger STX was only available as a 4x4 model but later in the model year the Ranger STX package was again offered in a 4x2 model as well.

Standard equipment found on the Ranger STX this year included the 140 horsepower rated 2.9 Liter EFI V-6 engine mated to a 5-speed manual/overdrive transmission, a limited slip rear performance axle; P215/75R15 SL white-letter-outlined, all-terrain steel-belted radial tires, and a special tuned suspension system.

Other standard equipment items found on the Ranger STX models this year included adjustable sport bucket seats with floor console, AM/FM stereo system, tachometer, leather wrapped steering wheel, Courtesy Light Group with headlights on warning buzzer, "Touch Drive Electric Shift" transfer case (4x4 models), black-colored bumpers, and unique "STX" sport tape exterior graphics.

And once again if a regular Ranger STX didn't look sporty enough an owner could pay extra and order an optional "Sport Appearance Package" in either a black or silver color. This "Sport Appearance Package" made the tough looking Ranger STX even tougher looking. This was a look that younger buyers seemed to find very appealing.

For those buyers who didn't want or need a sporty looking Ranger pickup, there were always the Ranger XLT models to look at, especially if they were looking for a deluxe pickup with some touches of luxury.

This XLT trimmed "Special Vehicle Package" came with the following equipment; a cloth-covered three passenger 60/40 split bench seat, a cloth-covered color-keyed headliner, color-keyed carpeting, power steering, a Courtesy Light Package, a tachometer, and a sliding rear window in Regular Cab models. The SuperCab XLTs came with most of those items plus a pair of rear-folding jump seats.

Both Ranger XLT versions came with a two-tone exterior paint job, chrome plated grille, chrome plated bumpers, an electronic AM/FM stereo cassette sound system; P215x14 inch white-letter-outlined, all-season radial tires for 4x2 trucks or P215x15 white-letter-outlined, all-terrain tires on 4x4 trucks. If you ordered your Ranger XLT with a manual transmission, Ford threw in a set of deep-dish cast aluminum wheels at no extra cost.

For those buyers looking for a lower initial cost pickup, there was the Ranger "S", Ranger "S Plus", and the Ranger Custom to choose from. All these lower cost Rangers came with a vinyl covered bench seat, a 2.3 Liter fuel injected inline 4-cylinder engine with a 5-speed manual overdrive transmission, an electronic AM radio with digital clock, and black-colored bumpers.

The Ranger "S" and the Ranger "S Plus" trucks were only available in a Regular Cab model, while the Ranger Custom was offered with a Regular Cab or a SuperCab. All three versions were available in either a 4x2 or a 4x4 layout.

For the 4x4 Ranger "S Plus" buyer, the Ford Motor Company offered either a regular steel pickup box or a new High Strength Composite (plastic) pickup box which was corrosion proof and at the same time constructed in such a way to resist denting and scratching.

The plastic pickup box exterior surfaces were formed by Reinforced Reaction Injection Molded (RRIM) polyuria plastic with interior surfaces made of polycarbonate plastic material. These plastic pickup boxes also used 31 fewer components compared to the regular steel beds found on Ranger trucks.

Other changes found on the 1990 Rangers included a standard black-colored grille for the STX, a new windshield molding, and a 20-gallon fuel tank for SuperCab models.

Another new feature found on the 4.0 Liter V-6 engine was a Data Communications Link that allowed technicians to monitor engine functions using a computerized diagnostic system.

At the end of the 1990 model year Ford reported

that they had produced 282,829 Rangers and that the most popular Ranger this year was the Ranger XLT.

On September 13, 1990 the Ford Motor Company introduced their new 1991 Rangers with a new promotional theme shared by other Ford trucks. This new promotional tag said "Ford Trucks-The Best Never Rest".Ford once again could claim that their Ranger was the "Number 1" selling compact truck in North America.

Leading the Ranger lineup this year was a new Regular Cab model that Ford called their "Ranger Sport". This new model came with all the regular equipment found on the Ranger Custom, Ranger "S",and Ranger "S Plus" plus cloth-covered bucket seats, tachometer, deep-dish cast aluminum wheels, and a unique sporty looking exterior tape striping.

The Ranger STX was still the-top-of-the-line model that came standard with the following items; a new 3.0 Liter/182 cubic inch multi port electronic fuel injection V-6 engine rated at 145 horsepower @ 4800 rpm for 4x2 models and a 2.9 Liter/179 cubic inch multi port fuel injection V-6 engine for 4x4 models; white-outlined-letter, 14-inch P215 all-season tires on 4x2 models and 15-inch P215, white-outlined-letter, all-terrain tires on 4x4 models.

Ranger STX trucks also came with cloth-covered sport bucket seats with power lumbar supports and power adjustable cushion bolsters for the driver's seat. They also came with speed control, an AM/FM stereo sound system with cassette player and digital clock, leather wrapped steering wheel, tachometer, "Touch Drive Electric Shift" on 4x4 trucks, Courtesy Light System with "headlights on" warning chime, fancy trimmed door panels, color-keyed carpeting, color-keyed headliners, fog lamps, and bumper guards.

A Ranger with a little touch of luxury was still the Ranger XLT and again the Ranger XLT would prove to be the most popular Ranger sold in 1991.

What made the Ranger XLT s so popular were the standard equipment pieces that made up the XLT package: items like a cloth-covered, three passenger 60/40 split bench seat, deluxe door panel trims with map pockets and lower areas covered in carpet, color-keyed carpeting, color-keyed headliner, AM/FM stereo system with cassette player, power steering, tachometer, and a Courtesy Light Package.

Other XLT items included exterior lower accent tape striping, chrome plated bumpers; P215x14,

A set of mis-matched wheels and factory applied tape graphics set this Ranger XLT Regular Cab pickup truck apart from the crowd.

Another 1991 Ranger XLT Regular Cab pickup truck with a set of optional wheels and graphic package.

white-outline-lettered, steel-belted all-season tires, steel-belted P215x15 white-outline-lettered, all-terrain tires on 4x4s, and all manual transmission equipped Ranger XLT s came with deep-dish cast aluminum wheels.

Roomier SuperCab models this year were available in Custom, XLT, or STX trim levels and all SuperCab models, except for those trucks with standard rear folding jump seats, came with a 22 cubic feet of storage space behind the front seat. Folding rear jump seats were standard equipment in XLT and STX trimmed trucks and optional equipment in Ranger Customs. And all Ranger SuperCabs this year still came standard with 20-gallon fuel tanks.

The Ranger Sport Preferred Equipment Package included a 60/40 split bench seat with a cloth cover, an electronic AM/FM stereo system with cassette

This Ranger XLT Regular Cab pickup truck has been personalized with a fancy graphic package, wheel well trim, side running boards, and a pickup truck bed cap.

player and digital clock combination, and a leather wrapped steering wheel.

Other Ranger Sport equipment included color-keyed sun visors, color-keyed carpeted floor mats, tachometer (4x2), power steering (4x2), unique exterior sport tape striping, deep-dish aluminum wheels, and P215 white-outline-lettered, steel-belted radial tires.

Ranger "S" and Ranger Customs were available in regular or long wheelbase models in 4x2 or 4x4 layouts and these Ranger models also came standard with 2.3 Liter inline 4- cylinder engines, 5-speed manual overdrive transmissions, power brakes, rear anti-lock braking systems, and all-vinyl trimmed interiors.

Ranger Customs and Ranger Sport models were available with optional 2.9 Liter V-6 engines, 3.0 Liter V-6 engines (4x2), and or 4.0 Liter V-6 engines with automatic overdrive transmissions.

There was a lot to brag about when it came to the 1991 Ranger lineup but the biggest news in Ford truck news this year was the release of a new sports utility vehicle called the Explorer which was built on a Ranger sized platform. That Explorer, in both two door and four door versions, would go on to become one of the most popular sport utility vehicles of all time. With the release of this new sport utility vehicle in late 1990, Ford didn't need the Bronco II in their truck lineup any longer, so the Bronco II quietly faded away at the end of the 1990 model year.

When Ford introduced their 1992 Ranger compact trucks in the fall of 1991 they promoted them as being both tougher as well as being sportier. Ross Roberts, Ford Vice President and Ford General Manager spoke about these new Rangers by saying

..."We keep making improvements because we realize that this is the only way to stay ahead of the pack in this high volume, extremely competitive market".

What made these 1992 Rangers tougher was Ford's improved corrosion protection program through their use of more two-sided galvanized steel body panels on all Rangers.

Sportier Ranger technical improvements included adding stabilizer bars to Ranger models equipped with power steering This helped make these Rangers ride and handle much better.

The optional 3.0 Liter V-6 engines received new roller tappets, sequential electronic fuel injection, and a single serpentine accessory belt system to replace the multi-belt accessory system.

Ranger 4x4 buyers still got the "Touch Drive Electric Shift" transfer case shifting system as standard equipment or they could choose a system with manual front hubs.

The 1992 Ranger lineup included the Ranger "S", Ranger Custom, Ranger Sport, Ranger XLT, Ranger STX, Ranger Custom SuperCab, Ranger XLT SuperCab, and the Ranger STX SuperCab.

For the Ranger Custom SuperCab there was a new "Comfort Cab" option for this model, an option that included a cloth-covered 60/40 split bench seat and other niceties.

In 1992 Ford called their Ranger Sport a "youthful machine", a model specifically designed to appeal to younger buyers. "Youthful-themed" equipment found on these models included a unique tape stripe treatment, white-outline-lettered, all-season steel-belted tires for 4x2 models and white-outline-lettered, all-terrain tires for 4x4 models and a set of deep-dish cast aluminum wheels too.

Other "Sport" equipment included a cloth-covered 60/40 split bench seat, electronic AM/FM stereo radio with cassette player, power steering, a leather wrapped steering wheel, tachometer, 2.3 Liter EFI inline 4-cylinder engine, and a 5-speed manual overdrive transmission.

To make that Ranger Sport even sportier a buyer could order a 3.6 Liter V-6 engine (4x2), a 2.9

Liter V-6 (4x4), and a 4.0 Liter V-6 with a 4-speed automatic overdrive transmission as options.

Ford referred to their 1992 Ranger STX trucks as their "dynamic" compact trucks. What made this truck so "dynamic" was all the sporty standard equipment that was found in this package. Sporty equipment like a tachometer, fog lamps, leather wrapped steering wheel, sport bucket seats, a floor console, new "STX" exterior tape striping, and P215x14 all-season, white-outline-lettered tires and P215x15 white-outline-lettered, steel-belted radial tires on 4x4 models.

Other Ranger changes seen during the 1992 model year was the deletion of a headliner from Ranger "S" and Ranger Custom models. With their headliners gone these models now featured painted interior top panels. Other changes for 1992 included an improved electric wiper switch and the addition of seat headrests for outboard passengers on the vinyl covered bench seats used in the Ranger "S" and Ranger Custom models.

This 4x4 Ranger XLT Regular Cab pickup truck features a nice chassis rake, a graphic package, and a hood protective molding.

1992 Ranger Exterior and Interior Colors

Exterior	Interior
Raven Black	Mocha, Scarlet Red, Medium Gray
Oxford White	Crystal Blue, Scarlet Red, Medium Gray
Cabernet Red	Scarlet Red, Medium Gray, Mocha
Bright Red	Scarlet Red, Medium Gray, Mocha
Mocha	Mocha
Medium Platinum Clearcoat	Scarlet Red, Medium Gray
Cayman Green Clearcoat Metallic	Medium Gray, Mocha
Wild Strawberry Clearcoat Metallic	Scarlet Red, Medium Gray, Mocha
Bright Calypso Green Clearcoat Metallic	Medium Gray
Mocha Frost Clearcoat Metallic	Mocha
Medium Mocha Clearcoat Metallic	Mocha
Brilliant Blue Clearcoat Metallic	Crystal Blue, Medium Gray
Twilight Blue Clearcoat Metallic	Crystal Blue, Medium Gray, Mocha (XLT)
Silver Clearcoat Metallic	Crystal blue, Scarlet Red, Medium Gray

This is what a nicely dressed up Ranger XLT Regular Cab pickup truck looked like back in 1990. (Ford Motor Company Photo).

Here are two photos of some sharp looking 1991 Ranger pickups. Upper photo shows an STX Regular Cab 4x4 version while the lower photo shows an XLT SuperCab model. (Ford of Canada Photo)

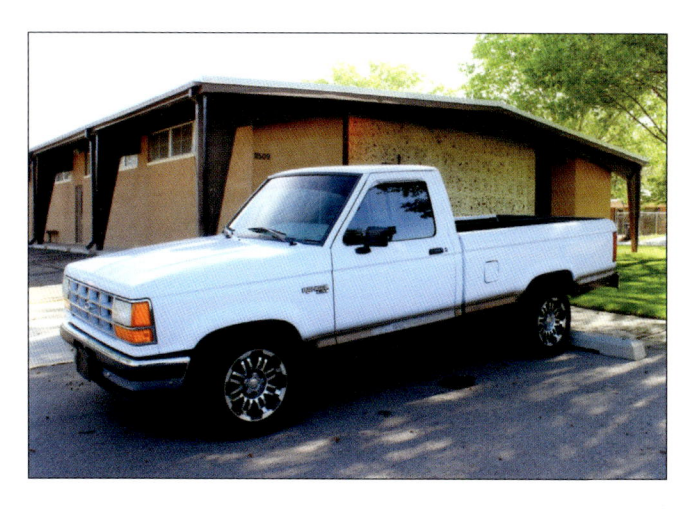

A set of modern chrome wheels dress up the looks of this 1992 Ranger XLT Regular Cab pickup truck.

A hood protective molding, a bed mounted tool box, and a tape graphic package are seen on this 1992 Ranger Regular Cab pickup truck.

"Where else can you have this much fun for under $10,000" was a good way for Ford to promote their new 1992 Ranger Regular Cab pickup trucks. This ad also shows one of the first times that Ford used their "Best Never Rest" campaign phrase.

53

Racing Rangers

Back in the 1980s and 1990s the Ford Motor Company was looking for ways they could promote the "toughness" of their Ranger trucks. Ranger toughness was a common theme in the company's advertisements and in some of their promotional materials. Besides talking about toughness the company decided to back up their toughness claims by actively supporting and providing some technical assistance to teams running Ford products in competitive events.

Two of the racing competitions that Ford Rangers competed in during that time were off-road or desert racing and sport truck racing. The former involved trucks running in races staged in desert areas like Baja while the latter had trucks racing on road courses where sports cars usually ran.

One of the major forces in off-road racing was the late Bill Stroppe who prepared successful racing vehicles for a number of different drivers over the years. When it came to building, racing, and winning in modified Ranger trucks in desert racing the team of Bill Stroppe and the late driver Manny Esquerra proved to be the team to beat whenever and wherever they showed up to compete. When asked about his success with racing Rangers Bill Stroppe was quoted as saying ..."I've been racing Fords for more than 30 years, and of all the vehicles I've prepared and modified for racing, the Ranger is the best. I could tell it was a winner when I first started putting it together for racing." While driving Stroppe prepared Rangers and Couriers, Manny Esquerra won a number of off-road races and eight or more SCORE Class 7 Compact Truck Championships.

If you have been around high performance Ford powered vehicles over the last thirty years or so, Steve Saleen's name should be familiar to you, especially if you like Mustangs. Steve Saleen was the driving force behind the Saleen Mustangs, some of the best limited-production high performance Mustangs ever created.

Besides building high performance Mustangs Steve has been involved in racing since the mid 1970s. During that time he has raced a number of different cars in Sports Car Club of America sanctioned professional events. In the late 1980s he built and successfully campaigned vehicles in two different SCCA Pro Road Racing Series.

In the SCCA Escort Endurance Series his team ran a couple of Mustangs while at the same time his other team was competing in the SCCA Coors Race Truck Challenge Series running a couple of modified Ranger trucks competing against Toyotas, Chevrolet S-10s, and Nissans. Both the Mustangs and Rangers proved to be worthy competitors and race and championship winners.

Both the Stroppe and Saleen race modified Ranger trucks proved beyond a shadow of a doubt that they were as tough and in some cases even tougher than the competition they faced.

Here we see a racing themed magazine ad showing one of the most successful teams in off-road racing. The Ford Ranger truck and the late Manny Esquerra who won quite a few championships driving Ranger trucks.

Still another racing themed magazine ad from the late 1980s. These ads promoted the Rangers as tough trucks.

Here we see one of the magazine ads touting the racing success of a couple of modified Ranger race trucks.

Another racing themed ad from the late 1980s showing two racing Ranger trucks.

This magazine story talks about Manny Esquerra, Bill Stroppe, the Ford Ranger, and Motorcraft parts that are used in their racing truck.

55

For 1993 Ford restyled their Ranger truck giving it a more muscular look. Here we see a Ford Motor Company promotional photo of a new Ranger SuperCab Styleside pickup truck.

Chapter 6: 1993-1997 The New Look Rangers Make Their Debut

On September 14, 1992, the redesigned Ranger compact pickup trucks that were built in Louisville, Kentucky, Edison, New Jersey, and the Twin Cities Plant in Saint Paul, Minnesota made their official debut.

These new 1993 Rangers didn't look anything like the Rangers they replaced. When the program to replace the original Rangers began in 1988, the Ford Truck Engineering Department set themselves a single goal… and that goal was to make the most popular compact pickup truck even better. When compared to the earlier Rangers, these new look Ranger trucks more than met that goal. These trucks were far superior to the competition they faced from the Chevrolet S-10, GMC Sonoma, Toyota's small pickups, the Dodge Dakota, and the Nissan Hard Body trucks.

Ford designers changed the looks of these new Rangers by making them 4 inches longer overall by lengthening their front ends and dropping the leading edge of their longer hoods by some 4 inches. At the same time, Ford engineers lowered the Ranger's chassis and enhanced the 1993 Ranger's riding comfort, road feel, and handling characteristics by widening the Ranger's track dimensions by 1.4 inches from 55.3 to 56.7 inches, and retuning the shock absorbers, suspension bushings, and spring rates to accommodate the new wider stance of these compact trucks.

Ford engineers also added rear stabilizer bars to the Ranger's standard equipment list. For Rangers with lower gross weight ratings, they replaced their steel rear leaf springs with new leaf springs made out of a composite fiberglass material to help save some weight. Ford engineers also decided to make some changes to the Ranger's steering system by redesigning the gear set up which provided for more precise steering and the system's self centering capabilities

A stiffer .217 inch diameter torsion bar helped to increase steering feedback to the driver plus more directional stability control. For those buyers ordering power steering, the control valve for the power steering system used on the 1993 models was modified in a way to better control fluid flow, which also resulted in a significant improvement in steering feel.

Other changes seen on the 1993 Ranger trucks included a redesigned, more car like oval muffler, wiper arms modified to give better contact between the wiper blades and the windshield, and the relocation of the EEC-IV control computer.

The computer control box was moved from inside the engine compartment to a new location inside the "A- pillar" area of the passenger side of the interior. The box itself, when mounted, was accessed through a small panel on the right side of the firewall. This one change better isolated the computer from heat, dirt, and moisture. Moving this computer unit allowed Ford engineers the chance to eliminate an 18-foot long wiring loom and also helped to give the engine compartment a less cluttered look.

Exterior changes included the curving of the sheet metal into a more cohesive aerodynamic design. Ford designers also added flush windshield and side window glass to cut down on wind resistance and noise. This also helped to make the body more aerodynamic. The designers also did away with the roof drip rails because they chose to use "limo" style doors.

Other design changes included new front and rear aero-styled headlamps and tail- lamps, a restyled tailgate with modified latches to make the tailgate easier to open and close, a new one piece contoured stamped steel rear bumper; a high-center roof-mounted stoplight at the rear of the cab, and a new grille design.,

All of these changes reduced the Ranger's wind noise, enhanced fuel mileage numbers, and gave the new 1993 Rangers a more modern look.

Not only did Ford stylists change the exterior looks of these compact trucks they also reworked the interiors to give them a more contemporary look as well. They made these interiors look more appealing by giving customers more colorful interior fabric choices. They also came up with eight different interior seat trim choices spread across four model lines.

Facing those seat arrangements was an instrument panel redesigned and rearranged to be more driver and passenger friendly with new graphics, relocated controls, and more eye-catching color-keyed control switches and buttons. Visibility, reachability, and

The "King of the "Thrill" was a clever way to introduce the restyled 1993 Ranger pickup truck line.

quick recognition controls were now easier to see, to reach, and to operate.

Other new interior items included new door trim panels with ergonomic controls, new steering wheels, and new seat trim with new sew stitch styles on all Rangers.

During the development stage for these new 1993 Rangers, the Ford Dealer Network let it be known to Ford executives that their customers wanted their 4x4 Rangers to look different than the Ranger 4x2 trucks. Ford stylists took their suggestions to heart and gave the 4x4 Rangers their own more muscular look. This distinctive new appearance was the result of a separate styled grille and a front valance panel not shared with Ranger 4x2 models. This front valance panel incorporated a set of unique round driving lamps. Four-wheel drive models also came with a set of P235/75R15L or P265/75R15 SL all-terrain, steel-belted radial tires.

4x4 Rangers also came equipped with a set of wheel

Here is an extensively customized California Ranger pickup truck that was done up this way to promote a truck magazine in the mid-1990s.

1993 Ranger Exterior Color Choices

Code	Description
EY	Vibrant Red (4x4 Only)
EG	Electric Red Clearcoat Metallic
GA	Dark Plum Clearcoat Metallic
PM	Bright Calypso Green Clearcoat Metallic
DA	Cayman Green Clearcoat Metallic
DD	Mocha Frost Clearcoat Metallic
DC	Medium Mocha Clearcoat Metallic
KE	Brilliant Blue Clearcoat Metallic
KN	Dark Blue Clearcoat Metallic
YO	Oxford White
YN	Silver Clearcoat Metallic
WC	Opal Grey Clearcoat Metallic
YC	Raven Black

1993 Ranger Interior Color Choices

Ruby Red   Royal Blue   Opal Grey

well flares in XLT, XLS, and STX models, and Ford's "Touch Drive Electric Shift" transfer case as standard equipment. Manual front locking hubs were also available as an option.

The new standard engine in the Ranger STX 4x4 trucks was Ford's 3.0 Liter EFI V-6 engine and 4x4 Rangers also came equipped now with a lighter duty Dana 28 front axle unless they were ordered with 4.0 Liter V-6 engines. In that case, the Ranger trucks still used the heavier duty Dana 35 front axle.

The 1993 Ranger model lineup included the base level Ranger XL which replaced the Ranger "S" and Ranger Custom. The XL was now a "no frills" model with a lot of extra cost options for buyers wishing to design their own fun trucks.

The Ranger XLS was the sporty version of the lower tiered models. This version came with all the standard equipment found on the Ranger XL plus a bold new tape stripe treatment that gave these trucks a sportier image.

The next trim level for the Ranger was the XLT, which Ford referred to as a luxury compact pickup truck for city or country living.

Then there was the Ranger STX, which was Ford's ultimate sport truck with large round fog lamps, wheel well moldings, and unique "STX" sport exterior tape graphics.

All of these models were available in short or long wheelbase models in 4x2 and 4x4 models in Regular Cab versions and SuperCab versions in XL, XLT, or STX trim levels.

All told, there were 22 models available to the Ranger buyer in 1993.

The new base model Ranger XL came standard with a vinyl covered bench seat, argent-colored styled-steel wheels, Ford's 2.3 Liter "Dual Plug" EFI inline 4-cylinder engine, 5-speed manual overdrive transmission, rear step bumper, and a floor mat in place of carpeting in the cab.

The Ranger Sport came equipped the same way plus an electronic AM/FM stereo radio/cassette player/clock combination, a sporty exterior tape stripe treatment, and argent-colored styled steel wheels or cast aluminum deep-dish wheels.

The Ranger XLT 's package was trimmed out quite a bit nicer than the lower tiered models with extra value items like a floor consolette, a sliding rear window, a 60/40 split bench seat, electronic AM/FM stereo radio/cassette player/clock combination entertainment system, cargo area lamp, power steering, and cast aluminum deep-dish wheels. These trucks also came with a set of chrome plated bumpers.

The "grand daddy" of the Ranger lineup was still the Ranger STX. This version came standard with a 3.0 Liter/145 horsepower rated EFI V-6 engine, a gray-colored rear step bumper, sporty cast aluminum wheels, a 4x2 handling package, special sporty "STX" tape striping, dual cloth-covered bucket seats with a floor console, black leather wrapped steering wheel, front valance mounted fog lamps, and other goodies shared with the Ranger XL, and Ranger XLT.

Prior to the release of the 1993 Rangers Ford

conducted some demographic studies that showed the median age of a Ranger buyer was 37 years of age and had a median income of over $40,000.00. And the mix of Ranger buyers showed that 84 % of them were males and 16 % were females. Thirty percent of them were college educated and seventy five per cent of them were married.

Through 1992, Ford had sold over 2.5 million Rangers. With these new look Rangers, Ford would sell even more of them.

..."Some see being the "best" as an end. At Ford we see it as the beginning. At the heart of our philosophy of continuous improvement is the notion that the best products can always be made better."

Those three sentences were used by the Ford Motor Company to introduce their 1994 Ranger compact trucks to the public in the fall of 1993. Those three sentences best described how the company was trying to make their products more appealing and keep them at the top of the sales charts..

Those improvements included refining the interior to make it more comfortable and appealing for both drivers and passengers. Another area of improvement was changes made to the suspension system, which resulted in Rangers getting a smoother and more controlled ride.

Making the interior look more appealing and smoothing out the ride were changes usually associated with improving the quality of passenger cars rather than trucks. By incorporating those changes in their Ranger trucks, Ford was able to market these trucks as being more versatile. They drove and handled like cars while maintaining the utility aspects of a truck.

The 1994 Ranger lineup included the Ranger XL, Ranger XL Sport, Ranger XLT, and the Ranger STX. Ford called the XL their rugged compact truck and their XL Sport a compact truck with a sporty flair. The next Ranger in the lineup was the value packed XLT, which Ford referred to as a "compact truck with a solid reputation for toughness and a long history of success. For anyone looking for a good truck with extra value, the Ranger XLT was a tough choice to beat as it came with standard interior features like a three passenger 60/40 split bench seat with a floor consolette/storage compartment, color-keyed carpeting, color-keyed cloth headliner, deluxe door trim panels with lower panel areas covered in carpeting and map pockets. Ranger XLT models also came with a special light package that included lights

Another customized Ranger SuperCab pickup making an appearance at a custom car show in the mid-1990s.

in the glove box, ash tray, engine compartment, and cargo areas. Other XLT equipment included a chrome plated front bumper, bright metal trim around the grille, a sliding rear window, power steering, and rear folding jump seats (SuperCab).

By adding the optional "XLT Preferred Equipment Package," the buyer got a truck with larger tires, an exterior accent paint stripe, an electronic AM/FM stereo radio with cassette player, and a chrome plated rear step bumper (standard on all 4x4 XLT trucks).

With all that extra equipment it was no wonder the Ranger XLT was the most popular Ranger model this year and other years as well.

In 1994, if one wanted a really dynamic looking pickup truck the Ranger STX was the top choice in compact trucks. The Ranger STX came standard with Ford's 3.0 Liter EFI V-6 engine bolted to a 5-speed manual overdrive transmission, cloth-covered Captains Chairs with a floor console (a set of sport bucket seats came standard with the STX Preferred Equipment Package), power steering, tachometer, distinctive "STX" exterior tape striping, and a leather wrapped sport steering wheel.

There was another Ranger model that made its debut in 1994 model year. This new model was called the "Ranger Splash" and it was a "youth-oriented vehicle with a uniquely designed "Flareside" style bed that combined a narrower bed with a wide set of rear fenders. It was the first time such a bed style was offered in the compact truck class. Now Ranger buyers could purchase a truck with a traditional slab-sided side bed or one with the narrower "Flareside" bed.

Close up shot of a 1994 Ranger XLT fender emblem.

AMT-Ertl released this plastic 1/25th scale model of a Ranger STX Regular Cab pickup truck in 1994. It was also available in a red color.

step bumper, color-keyed grille surround trim, color-keyed front valance cover, and deep-dish aluminum wheels.

The 1994 Ranger story wasn't all about comfort and power because Ford did add an extra measure of safety to the Ranger interiors by adding steel safety beams in the doors to help minimize the potential of a cab intrusion in the case of an accident.

As part of their sales training for the 1994 Rangers Ford salesmen were given promotional materials comparing their Ranger models against different models from Toyota, Dodge, Nissan, and Chevrolet. These sales promotional materials showed the Ranger either ahead by a wide margin or even with the competition in most categories.

With 324,352 Rangers built in 1994 the Ford sales team learned their lessons well and did a good job of promoting these Ranger trucks to potential customers.

Following their claim to be always trying to improve the breed the Ford Motor Company made a number of significant changes to their 1995 Ranger models.

One of those changes involved installing a driver's side air bag as a piece of standard equipment. Another safety related change was to add a 4-wheel anti-lock braking system under all 4x4 Rangers this year. There was also a brake/shift interlock system built into the automatic transmission offered as an option on Ranger trucks.

A new instrument panel design with improved instrumentation and new controls was added. Other interior changes included the Ranger's electronic speed control system which was redesigned to make it easier to use, a new 6-disc CD changer that replaced the Compact Disc radio option found in earlier Rangers, and the addition of a new 6-way power driver's seat that made adjusting the seat easier, and height adjustable shoulder harnesses that allowed the driver and outside seat passengers to be more comfortable and safer at the same time.

For those Ranger owners who were "gadget freaks" Ford installed a power point outlet on the instrument panel to power up accessories.

For 1995 the Ranger lineup once again included the Ranger XL, Ranger XL Sport, the Ranger XLT, the Ranger STX, and the Ranger Splash.

All those improvements helped to boost Ranger production for the 1995 model year to over 356,700 units.

The Splash was available in a Regular Cab or in a Super Cab version as a 4x2 or a 4x4 model. The Splash 4x2 version was a solid looking vehicle and with its low-riding stance as an eye-catching, head-turning, sporty truck. That low stance combined with a special handling suspension system, and 15-inch wheels produced a very confident handling machine that was a lot of fun to drive.

Ranger Splash 4x4 vehicles on the other hand had a "macho" look and with a standard 3.0 Liter EFI V-6 multi-port, fuel injected engine under its hood had the power to back up those looks.

Other Splash goodies included "Splash" exterior tape striping, color-keyed exterior power mirrors, color-keyed front bumper cover, color-keyed rear

"It's stylish and lots of fun to drive. Tough and versatile too". These are the terms Ford used to describe their 1996 Ranger models when they were introduced in the fall of 1995.

For 1996 the Ford Ranger was the first compact pickup truck to offer buyers a passenger side air bag as a piece of optional equipment. All 1996 Rangers also came with an illuminated entry system that turned on the interior lights when the driver's door handle was operated. A remote keyless entry/anti theft system was made optional in 1996 as well and this system could be ordered on Ranger XLT, Ranger STX, and Ranger Splash models.

Also new this year as an option on SuperCab Ranger XLT models was a set of pivoting rear quarter windows In addition, the sliding rear window for the Ranger XLT, Ranger STX, and Ranger Splash models now featured "Privacy Glass".

Another change involved Ford adding a 4.10:1 rear axle to all Ranger models this year equipped with P235 tires.

For the 1997 model year, the Ranger was promoted as being "The Foremost Authority On Being Fun Tough" and also promoted as "One Very Tough Customer".

New for 1997 was to offer as an option on the Ranger XLT models power windows and power door locks. For those buyers who liked the Flareside pickup bed used on the "Splash" model, this particular pickup box or bed was now an option for all 4x4 vehicles. Ford also made the 3.0 Liter EFI V-6 engine, standard equipment for all Ranger 4x4 vehicles.

The Ranger Splash now featured new body-colored exterior molding, a retuned suspension package, and a sliding rear cab window.

Ranger XLT, Splash, and STX models featured redesigned exterior tape stripe graphics. Rangers equipped with 4.0 Liter V-6 engines could be ordered with a new Ford 5R55E, 5-speed automatic overdrive transmission as an extra cost option. And the Ranger's 4R44E 4-speed automatic overdrive transmission was modified and upgraded this year with some of the tougher components used in the 5R55E transmission.

A second 12-volt power point outlet was added to the Ranger's 1997 instrument panel and on models not equipped with a passenger side air bag, another 12 volt power point outlet was added to the mix.

At the time this photograph was taken, this Ranger XLT SuperCab pickup truck was over twenty years old. It spends a lot of time in a garage and still looks factory fresh.

This 1995 Ranger XLT SuperCab pickup is painted in a color Ford called "Brilliant Blue Clearcoat Metallic".

This Ranger custom model was a snap together kit made for young car enthusiasts.

Exterior Paint Colors with Interior Color Options

Cayman Green Clearcoat Metallic	Medium Graphite, Saddle
Mocha Frost Clearcoat Metallic	Saddle
Bright Red Clearcoat	Medium Graphite, Saddle
Electric Red Clearcoat Metallic	Medium Graphite, Saddle
Sapphire Blue Clearcoat Metallic	Medium Graphite, Saddle
Dark Blue Clearcoat Metallic	Royal Blue, Medium Graphite
Medium Willow Green Clearcoat Metallic	Willow Green, Medium Graphite, Saddle
Charcoal Grey Clearcoat Metallic	Royal Blue, Medium Graphite
Black Clearcoat	Medium Graphite, Saddle
Silver Clearcoat Metallic	Royal Blue, Medium Graphite
Oxford White Clearcoat	Royal Blue, Willow Green, Medium Graphite, Saddle
Canary Yellow Clearcoat (Splash Only)	Medium Graphite, Saddle

Last but not least, three new exterior color choices were added to the 1997 Ranger color palette. These new colors were Light Prairie Tan Clearcoat Metallic, Toreador Red Clearcoat Metallic, and Portofino Blue Clearcoat Metallic.

At year's end, Ford Ranger pickups were again the "Number 1" selling vehicle in the compact truck segment.

A Ranger SuperCab pickup truck sits on top of a car hauler with other new Ford trucks on their way to car dealers.

Another new Ranger XLT Regular Cab pickup truck sits on the upper tier of a car hauler in transit to a Ford dealership.

In the late 1990s scenes like this one showing a line up of new Ranger pickups was a common sight on Ford dealer lots across this country.

Here we see a really sharp looking Ranger Regular Cab pickup truck in a Toreador Red color.

This Ranger XLT SuperCab pickup truck has been equipped with side boards and a set of factory mud flaps.

This drag racing modified Ranger pickup truck belonged to noted drag racer Roy Hill and he displayed it at the "100th Anniversary of Ford Racing" in Dearborn, Michigan back in 2001.

A couple of older Ranger pickups sit on a used car lot in Albuquerque in 2015.

Ranger pickup trucks like this XLT Regular Cab version in white have always been popular with business owners because they are good looking trucks that are relatively economical to operate.

A sharp looking Ranger Regular Cab 4x2 pickup truck sits atop a car hauler on its way to a new owner.

You don't see something like this everyday... a Toyota pickup truck with a Ford oval grille emblem and Ford oval emblem wheel covers.

Here we see an off-road racing Ranger truck modified with wide fenders, driving lamps, tubular bumpers, extra skid plates, and extra shock absorbers.

This Ranger Splash is done up in a high profile Canary Yellow Clearcoat color with a pickup bed cap in the same color.

Here we see a sharp looking Ranger XLT 4x2 SuperCab pickup in a light tan color nicely highlighted by grey colored wheel arches.

This Ranger XLT SuperCab 4x4 model is finished in a two-tone red over black color combination.

Ranger XL Regular Cab 4x2 pickup truck with 6-foot Styleside bed in an Oxford White color.

Some Ranger Splash models like this one are painted in high profile colors like this red Clearcoat color.

67

This Ranger Splash model features a fancy looking bed cap painted in the same color.

This side view comparison photo shows the size difference between a Regular Cab XL Ranger pickup parked side by side with a larger F-150 Crew Cab pickup truck behind it.

1997 Ranger showroom catalog cover showing a Regular Cab and a SuperCab version along with some young people engaged in sport activities.

This 1998 Ranger XLT SuperCab 4x2 with Styleside bed is painted in a Pacific Green Clearcoat Metallic color. It is also equipped with mud flaps and a set of cast aluminum wheels.

Chapter 7: 1998-2000 Making A Good Truck Even Better And More Appealing

1998 was definitely a year of multiple changes for Ford's popular Ranger pickups and one of those changes involved the Ranger STX models; models that had carried the "Sporty" banner for the Ranger line since 1986. For whatever reason, Ford thought they no longer needed the STX in the Ranger lineup, so this model was dropped for 1998.

The 1998 Ranger lineup now included the Ranger XL as the base model, the Ranger XLT as the deluxe model, and the Ranger Splash as their youth-oriented sports model.

The Ranger was available in Regular Cab or SuperCab model in 2-door and 4-door versions, in short or long bed models, in 4x2 or 4x4 chassis forms. For those Ranger buyers who wanted a Regular Cab version, Ford redesigned the cab by extending it some 3 inches to 107.3 inches, giving these cabs more storage room as well as providing more room for seat travel and tilt.

Other new features found on the 1998 Rangers include new tow hooks front and rear for 4x4 models, a new front crush zone that helped to absorb more energy in front end collisions, a new 2.5 Liter inline 4-cylinder EFI engine rated at 119 horsepower at 5000 rpm with a maximum torque rating of 146 lb/ft at 3000 rpm, and a boxed frame for added strength. Then there was the redesigned and re-engineered front suspension system that incorporated short and long arm components, plus torsion bars which improved both ride quality along with handling characteristics.

For 4x4 Ranger buyers a new "Pulse Vacuum Hublock" (PVH) "shift on the fly" system made it easier and smoother to shift into and out of four-wheel drive. And, as an added bonus, the driver was not required to back up his Ranger to disconnect the front hubs as had been done on the earlier electric system. With this new system with (PVH) front hubs, those hubs disengaged all the four-wheel drive driveline components when the driver shifted out of that position.

To make it easier to get into and out of the XLT and Splash 4x4 equipped models, Ford added some interior "A" pillar mounted grab bars. Once inside these new cabs the driver and passengers sat on a redesigned, more comfortable seat. To make these cabs safer, Ford fitted these new Rangers with new second-generation de-powered driver and passenger air bags. The air bag on the passenger side could be deactivated by a manual switch if needed.

To make these new Rangers easier to steer and control, Ford outfitted them all with a new power rack and pinion steering system which was quite an improvement compared to the earlier systems used on the Rangers.

All Rangers were equipped with rear wheel ABS systems and four-wheel ABS systems were available as optional equipment. To increase corrosion resistance in these new Ranger models, Ford installed stainless steel exhaust systems and aluminum hoods on all of them.

For improved traction and handling, the 4x2 Rangers were fitted with larger tires as standard equipment. This year Ranger buyers were offered the choice of seven different tires to meet a wide variety of needs. In addition to a wider selection of tires, Ford offered 4x2 Ranger buyers a choice of four different wheels and four additional wheel designs for 4x4 models.

In order to accommodate a longer Regular Cab and still retain the same pickup bed lengths for these models, Ford engineers stretched the wheelbase of these trucks by some 3.6 inches to 111.6 inches for the short bed models and 117.6 inches for the long bed models.

Let's look at some of the improvements for this model year. Besides the newly redesigned aluminum hoods previously fitted to these Ranger trucks, the new front-end look included redesigned, aerodynamic halogen headlamps; a redesigned grille, a redesigned front bumper, and modified front fenders. On the rear, Rangers were fitted with redesigned tri-color tail lamps.

A couple of Ranger pickups sit atop a car hauler on their way to a Ford dealership somewhere in the USA.

Interior changes included newly designed seats that covered with new, richer looking fabrics. Instrument panel lighting was also changed to make the gauges easier to read. By stretching out the Regular Cab, an extra 4.0 cubic feet of storage space was found behind the front seat.

Ranger XLT and Splash models were now equipped with a new 4-pin towing harness for a more convenient way to make an electric connection for a trailer. XLT and Splash models now came with body-colored door and tailgate handles. All Ranger Regular Cab models were also fitted with larger rear windows.

Ranger XL models came with a grille with painted trim while the Ranger XLT models came standard with a grille with chrome plated accent trim. The Ranger Splash still came with a special textured grille with body-colored trim items. The bumpers on the Ranger XLS were fitted with painted bumpers, the Ranger XLT models came with chrome plated bumpers, and the Ranger Splash models once again came with color-keyed front and rear bumpers.

Another new feature in Regular Cab XLTs and Ranger Splash models this year was a behind the seat storage tray. The spare tire mounted below the pickup beds on Rangers was now easier to remove thanks to a new crank and cable security system mounted to the frame.

Self-adjusting clutches and modified 5-speed manual overdrive transmissions were now easier to use and new larger radiators helped the Rangers to run cooler and be more efficient.

The floor console that came with the Ranger's sport

At a car show in California we ran across this 1999 Ranger Regular Cab EV electric vehicle in very good original condition.

This sharp looking Ranger XLT Regular Cab 4x2 pickup truck with Styleside bed is finished in a nice two-tone Oxford White over Light Prairie Tan Clearcoat color combination.

bucket seats now had provisions for storing CDs, cassettes, coins, and a couple of beverage containers.

Ford's 3.0 Liter Vulcan V-6 EFI engine was modified with a new upper manifold assembly, which increased air flow which in turn helped to boost this engine's torque rating by 14 % to 185 lb/ft at 3750 rpm.

1998 Ranger Exterior Color Choices

Light Prairie Tan Clearcoat Metallic
Deep Emerald Green Clearcoat Metallic
Bright Atlantic Blue Clearcoat Metallic
Pacific Green Clearcoat Metallic
Light Denim Blue Clearcoat Metallic
Oxford White Clearcoat
Bright Red Clearcoat
Toreador Red Clearcoat Metallic
Medium Platinum Clearcoat Metallic
Boysenberry Blue Clearcoat Metallic
Black Clearcoat
Autumn Orange Clearcoat Metallic

SuperCab Ranger models this year were available with two or four doors. This four-door SuperCab option added about $700.00 to the base price of a SuperCab this option made it easier to access the rear storage compartment from both sides of the vehicle. These optional rear doors were hinged off the rear corner posts and they were able to swing out a full ninety degrees from the door sills eliminating the need for a "B" pillars that could obstruct loading or unloading of the vehicle.

Sometime during the 1998 model year the Ford Motor Company added an electric powered vehicle to the Ranger lineup. The first mention of this electric vehicle was a report of some of these first Ranger Electric (EV) vehicles being delivered to California Edison for testing in February of 1998. These electric vehicles were powered by a 90 horsepower, Siemens 3-phase electric motor, that was mounted in the rear underside of the vehicle. These Rangers looked just like regular gasoline engine vehicles unless one knew what to look for.

The first change that stood out was a little door that sat on the right hand side of the grille, a door that covered a receptacle where you plugged in a charging cord. Other changes of note were a special one speed 3:1 geared automatic transmission that was controlled by a transmission selector that showed positions for "Park", "Reverse", "Neutral", "Drive", and "Economy".

The owners manuals that came with these vehicles suggested that the "Drive" position only be used when the vehicle was driven on the freeways and the "Economy" position only be used in stop and go city driving situations because the power output in this position was less than what was available when the pickup was in the "Drive" position.

This Ranger EV model was only available in a Regular Cab 4x2 XL trimmed vehicle. These early Ranger EV vehicles were powered by a set of 8 volt

Delphi lead-acid batteries that were mounted in a box under the rear of the vehicle where the spare tire was normally carried, forcing the spare tire when it was carried in the vehicle to be moved into the bed of the truck. The beds on these trucks were also fixed with a cover to cut down on power losses. These Ranger EV vehicles were not sold to the general public, instead they were leased out to commercial fleets or business owners.

In all, these redesigned 1998 Ranger models offered the compact truck buyer more "bang for his buck" and helped keep the Ranger the number one seller yet again.

With all the changes made on the 1998 Rangers there weren't any major changes planned for the 1999 Ranger models that were introduced in the fall of 1998. Most of the changes that were made had to do with equipment being moved from the optional list to a standard equipment Ranger list.

On change that didn't have anything to do with equipment changes was the dropping of the Ranger Splash model. Ford replaced this youth-oriented model with a "Sport Appearance Group Option" for the Ranger XLT. This Ranger XLT with "Sport Appearance Group Option" now carried the same equipment that the separate "Splash" model came with previously: items like body-colored grille surround trim, body-colored front and rear bumpers, body-colored wheel lip moldings, and body-colored exterior mirrors.

For those Ranger buyers who liked to go off-road with their vehicles Ford offered them a new optional "Off Road Package" which consisted of a 4.10:1 rear axle; P245/75R 16-inch, white-outline-lettered, all-terrain tires; chrome plated grille and bumpers, and bright metal special exterior mirrors.

Air conditioning was now part of the Ranger XLT package but the Ranger XLT's body side stripe kit and two-tone paint treatments were now dropped. Another feature that was now part of the standard equipment package of the Ranger XLTs was a four wheel ABS braking system as were an AM/FM cassette radio, 15-inch cast aluminum wheels, and a standard spare tire lock.

Ranger XLT models with V-6 engines also came standard with Ford's SecuriLock Security System which helped to minimize the potential for these Rangers to be stolen.

Ford knocked five metallic color choices off the

Another car hauler carrying a load of trucks including a Ranger pickup out to the West Coast.

Three quarter rear shot of a Ranger SuperCab 4x4 pickup truck showing its sporty looking "Flareside" pickup bed.

Ranger exterior color list this year. Those deleted colors were Light Prairie Tan, Boysenberry Blue, Light Denim Blue, Deep Emerald Green, and Pacific Green. Ford did add some new color choices for the Ranger this year and those colors were Amazon Green, Deep Wedgewood Blue, Harvest Gold, and Jalapeno Green.

A new interior color of Dark Graphite was introduced for Rangers this year while the previously offered interior colors of Willow Green and Denim Blue were dropped. Besides Dark Graphite the other interior colors available were Medium Prairie Tan, and Medium Graphite.

All 1999 model year Rangers were now rated as "Flexible Fuel Vehicles", or FFVs. Flexible fuel meaning that they could be run on 100% regular unleaded fuel or E85 ethanol fuel, and any combination of the two fuels. For those Ranger buyers who wanted to lower their carbon footprint even more, Ford once again offered the Ranger EV

This is a close up detailed shot of "Ranger XL" front fender emblem.

vehicles in the Ranger lineup. Once again these electric vehicles were offered through a lease program to mostly fleet buyers. According to the Southern California Edison report in late 1998, several of the Ranger EVs they tested were new equipped with NiMH batteries. While these tests were being conducted, all the Ranger EV vehicles that were in fleet use were being powered by the old lead acid batteries.

Ford promoted these electric Rangers as being "Dependable and Reliable along with being Built Ford Tough". They went further by stating that "The nickel-metal hydride (NiMH) powered Ranger EV is a practical, robust, light duty pickup that balances performance, reliability, and energy efficiency".

From our research we have found that these Ranger EV trucks were all painted in an Oxford White Clearcoat exterior color with gray-colored interior trimmings. They were all trimmed out as Ranger XL 4x2 Regular Cab short wheelbase models that were mounted on Ranger 4x4 chassis.

These Ranger XL EV models looked the same as regular gasoline engine models except for the fact that they didn't have or need a tail pipe or an exhaust system because they were zero emission vehicles. They also still had the little door on the right side of the grille that, when opened, allowed a charging cord to be plugged in.

Another clue that these Rangers were not your regular gasoline fueled trucks was the "Electric" script identifier that was mounted just below the Ranger emblem that was installed on the front fenders.

With its hood open, you would see the charger input unit for the batteries, an electric air conditioning unit, a power steering set up, a power brake unit, a radiator for the air conditioner, a reservoir for the windshield washers, and a vacuum pump and reservoir for the power brake fluid.

The Alternating Current Controller was mounted under the pickup bed at the rear of the frame near where the spare tire was usually located and the spare tire was moved into the bed to make room. The rear wheels were powered by a six-pole alternating current motor operating through a single speed 3 to 1 ratio reduction automatic transmission and differential unit. This motor unit was mounted high up in the frame and the power it produced was then transferred to the wheels by a set of half shafts that were angled downward to a deDion style rear axle set up that was

Front 3/4 shot of a 2000 Ranger SuperCab 4x4 pickup with sporty "Flareside" pickup bed.

suspended on a pair of longitudinal leaf springs.

The first Ranger EVs used lightweight carbon fiber monolithic rear leaf springs but it was found that this set up had problems controlling the lateral movement of the rear axle.

This early suspension system also used a Watts link set up to minimize suspension and axle movement. But when Ford engineers replaced those carbon fiber rear springs with standard steel springs the problem of lateral axle movement disappeared and they also discovered they didn't need to use a Watts link to control this problem.

Ranger EVs also came with a set of low rolling resistance radial tires mounted on 15- inch cast aluminum wheels. With a bed cover plus those special tires, these electric powered trucks could score some impressive mileage figures.

In order to power these electric motored Rangers, a set of onboard batteries were needed and those batteries whether they were the old style lead-acid units or the newer (NiMH) type were carried in one big box that was attached to the underside of the vehicle. This box was removable for servicing with the use of some special designed shop tools.

Once again unless you were up close to these vehicles they looked pretty much the same as a regular Ranger pickup of the day. They even looked the same on the interior except for the gauges, indicator lights, and shift quadrant markings of "Park", "Reverse", "Neutral", "Drive", and "Economy".

These Ranger EV vehicles used regular speedometers and ammeters but in the instrument panel where a tachometer was used on gasoline models there was "a Miles to Go Gauge" and a "Charge" indicator replaced the normal fuel gauge. There was also an "Off-Run" electric gauge, which showed "Run" after the vehicle was started.

Just because these vehicles were dressed out as base Ranger XL models, didn't mean they didn't come equipped with some comfort and convenience items. Items like an AM/FM cassette stereo radio, power steering, air conditioning and a heater and defroster. Another piece of standard equipment on these trucks was a vinyl bed cover to help cut down on aerodynamic resistance.

The electric Rangers were powered by a 90 horsepower rated Siemens 3 Phase electric motor, that also carried a maximum torque rating of 140 lb/ft at a low rpm rate. As for charging, the Ranger EV

A Ranger SuperCab XLT pickup with sporty looking "Flareside" pickup bed sits on top of a car hauler heading to a new owner.

Though this is the base model Ranger Regular Cab with a short 6 foot bed it still is a nice looking truck for any business or private owner.

vehicles used a 240V/30 amp recharging system and it took anywhere from 6 to 8 hours to fully recharge the batteries.

1999 Ranger Base Prices	
XL Regular Cab SWB (short wheel base)	$11,845.00
XL Regular Cab LWB (long wheel base)	$12,315.00
XLT Regular Cab SWB	$13,920.00
XLT Regular Cab LWB	$14,470.00
XL SuperCab	$15,300.00
XL Regular Cab 4x4 SWB	$16,195.00
XL SuperCab 4x4	$17,335.00
XL Regular Cab 4x4 LWB	$16,705.00
XLT Regular Cab 4x4 SWB	$17,735.00
XLT Regular Cab 4x4 LWB	$18,310.00
XLT SuperCab 4x4	$19,435.00
XL Ranger EV (Lease)	$24,990.00

Businesses like Century Link found that Ranger pickup trucks, like this SuperCab model, make for good economical work trucks that are very reliable.

Another base model Ranger Regular Cab 4x2 pickup truck with 6 foot Styleside bed.

Ranger XLT SuperCab 4x4 pickups like this one have always been popular and are a common sight today on our highways and byways.

Rear 3/4 shot of a Ranger XLT SuperCab 4x4 pickup with aftermarket "headache" rack and upper body bed rails.

76

1999 Ranger Estimated Fuel Economy Numbers

FFV 4x2 3.0 Liter EFI V-6 Engine with Automatic Transmission
    15 City   20 Highway   17 Combined

FFV 4x2 3.0 Liter EFI V-6 Engine with Standard Transmission
    16 City   21 Highway   18 Combined

4x2 2.5 Liter 4-cylinder Engine with 4 Speed Automatic Transmission
    18 City   23 Highway   20 Combined

4x2 2.5 Liter 4-cylinder Engine with 5-speed Manual Transmission
    19 City   24 Highway   21 Combined

4x2 4.0 Liter EFI V-6 Engine with 4 Speed Automatic Transmission
    14 City   20 Highway   16 Combined

4x2 4.0 Liter EFI V-6 Engine with 5-speed Manual Transmission
    16 City   20 Highway   17 Combined

FFV 3.0 Liter EFI V-6 Engine 4x4 with 4 Speed Automatic Transmission
    14 City   18 Highway   15 Combined

FFV 3.0 Liter EFI V-6 Engine 4x4 with 5-speed Manual Transmission
    14 City   18 Highway   16 Combined

4.0 Liter EFI V-6 Engine 4x4 with 5-speed Manual Transmission
    15 City   19 Highway   16 Combined

Ranger EV Electric Vehicles

4x2 with 4 Speed Automatic Transmissions
    54 City   63 Highway   58 Combined

MPGE Rating
    62 City   54 Highway   58 Combined

KWH/100 Miles
    69 City   72 Highway   72 Combined

Even though there weren't any real major changes to the Ranger for the 1999 model year Ford sold 348,358 Rangers that year, or 32.2 per cent of the entire compact truck mrket.

For the 2000 model year, Ford offered their Ranger pickup trucks in XL (base) and XLT (deluxe) trims. They also offered them in 4x2 or 4x4 chassis layouts, in Regular Cab or SuperCab modcls in 2-door or 4-door versions, with 4-cylinder or V-6 engines with manual overdrive transmissions or optional automatic overdrive transmissions, and with regular unleaded gasoline engines, Flexible Fuel Vehicles, and zero emissions electric motored vehicles.

Ford also offered Ranger pickup trucks with short wide beds, long wide beds, and pickup beds in either Styleside or Flareside styles.

The Ranger engine lineup included a 2.5 Liter inline 4-cylinder engine rated at 119 horsepower, a 3.0 Liter V-6 engine rated at 145 horsepower, a 4.0 Liter V-6 engine rated at 160 horsepower, and a 90 horsepower electric motor.

Those engines and electric motor vehicles were backed up by a 5-speed manual overdrive transmissions, or optional 4-speed automatic overdrive or 5-speed automatic overdrive automatic transmissions, or the 1-speed automatic transmission found behind the electric motor powered Rangers.

The base model Ranger XL came with black-colored bumpers, grille, and exterior items, along with vinyl covered seats, a black vinyl floor mat, an AM/FM stereo radio, and vinyl-covered door panels.

The deluxe trimmed Ranger XLT models came with deluxe door panels, chrome plated grille trim, an AM/FM/CD stereo sound system, cloth-covered seats and more.

This Ranger SuperCab 4x2 pickup truck has been customized with a set of late model custom wheels.

A copper colored Ranger XLT SuperCab 4x4 model with Flareside styled bed sits on top of a car hauler in transit to a new owner.

Like previous model years 4x2 Rangers came with a coil spring style front suspension system while the 4x4 Rangers came with a torsion bar styled front suspension system which made them sit a bit higher. For 2000 Ranger buyers who were looking to buy a 4x2 pickup with the "macho" looks of a 4x4 vehicle, Ford offered a new optional "Off-Road Package" or "Trailhead Group Option" for Ranger 4x2 trucks.

This "Off-Road Package" had Ford putting on a 16-inch set of 5-spoke cast aluminum wheels with all-terrain tires for extra ground clearance. That had these 4x2 trucks sitting as high as the Ranger 4x4 vehicles. Two-wheel drive Rangers equipped with this option also featured the Ranger 4x4 front torsion bar suspension system rather than the normal front coil spring suspension system found on 4x2 Rangers not equipped with this option.

A set of tow hooks and argent-colored grille and bumpers added to that "Macho" look.

2000 Ranger Base Prices

XL 4x2 Regular Cab SWB	$11,580.00
XL 4x2 Regular Cab LWB	$12,050.00
XLT 4x2 Regular Cab SWB	$13,655.00
XLT 4x2 Regular Cab LWB	$14,205.00
XL 4x2 SuperCab	$15,240.00
XLT 4x2 SuperCab	$15,890.00
XL 4x4 Regular Cab SWB	$15,920.00
XL 4x4 Regular Cab LWB	$16,390.00
XL 4x4 SuperCab	$17,530.00
XLT 4x4 Regular Cab SWB	$18,070.00
XLT 4x4 Regular Cab LWB	$18,640.00
XLT 4x4 SuperCab	$19,785.00
XL Ranger EV (Lead Acid Batteries)	$35,100.00
XL Ranger EV (NiMH Batteries)	$49,105.00

2000 Ranger Popular Options

5R55E Ford 5-speed Automatic Overdrive Transmission (Only Available with 4.0 Liter V-6 Engine)	$1145.00
Limited Slip Rear Axle	$295.00
4.0 Liter EFI V-6 Engine	$695.00
Power Equipment Group (Power locks, Windows, Remote Keyless Entry)	$535.00
XLT Sport Appearance Group Package	$395.00
Air Conditioning	$805.00
4 Door SuperCab Option	$700.00

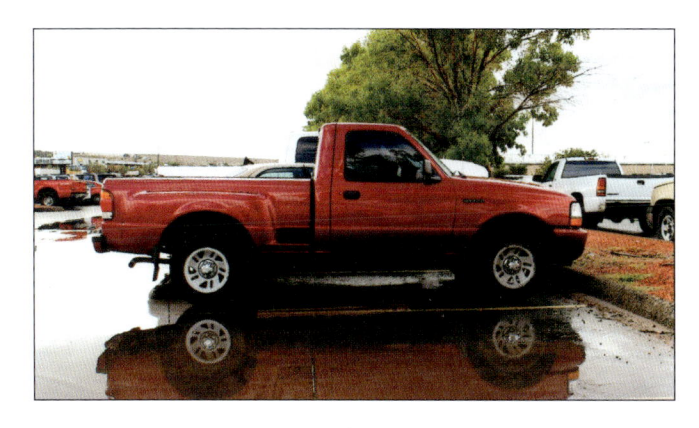

This sharp looking red colored Ranger Regular Cab XLT model sits with its reflection in a big puddle in a storage lot in 2015.

The graphics package, Flareside pickup bed, and chrome plated wheels really dress up the looks of this Ranger XLT SuperCab truck.

For those who wanted a deluxe-equipped, sporty looking truck; the Ranger XLT with the "Sport Appearance Group Package" would have been the one to choose. In picking this XLT option, the buyer got the usual deluxe equipment found on XLT models plus fog lamps and 4x4 XLT buyers got the fog lamps plus a skid plate mounted under the front axle.

Ford called these 2000 Rangers as being..."their tough trucks in a smaller size" and that was no idle boast. These trucks fit that description to a "T". Compact truck buyers must have liked that "toughness" promotional angle because they bought more than 300,000 Rangers by the end of the model year.

A blue colored Ranger Regular Cab Flareside 4x4 pickup adorns the cover of this 1998 Ranger showroom catalog.

This Ranger Regular Cab pickup has been dressed up with bright metal wheel well trim, Mylar style lower body trim, and upper body bed trim pieces.

Chapter 8: 2001-2003 More Power, More Comfort, and More Changes For The Ranger

Since the Ranger was introduced back in the early 1980's Ford's goal had always been to continue to improve it and to make it more appealing to buyers each and every year. That is what potential buyers saw in late summer of 2000 when an improved 2001 Ranger lineup made its debut.

The Ford Motor Company promoted these new Rangers as offering more power, more performance, more personality, and a level of refinement not seen in the compact truck end of the market.

On the outside, these new 2001 Rangers featured a bold, aggressive styling that included changes to the grille area, bumpers, and restyled headlamps and side marker lights. In front of that new restyled hood sat a new look mesh type grille or a new two-bar chrome plated grille that made these two models stand out from the rest of the Ranger line. The Ranger XLT trimmed 4x4 vehicles and a new "Edge" model got an aggressive looking power-dome hood and revised wheel lip moldings that gave these trucks more "street credit", a look that was supposedly what young compact drivers and owners were looking for.

A Ranger SuperCab 4x4 pickup sits in an auction storage yard in 2015 waiting for its turn to go across the auction block to a new home.

2001 Ranger Base Prices

XL	Regular Cab	2WD	SWB (Short Wide Bed)		$12,110.00
XL	Regular Cab	2WD	SWB	3.0 Liter V-6 Engine	$12,505.00
XLT	Regular Cab	2WD	SWB		$14,220.00
XLT	Regular Cab	2WD	Appearance Package		$14,445.00
XLT	Regular Cab	2WD	Flareside Bed		$14,595.00
XL	Regular Cab	4WD	SWB	3.0 Liter V-6 Engine	$14,700.00
XLT	Regular Cab	4WD	Appearance Package		$14,920.00
XLT	Regular Cab	4WD	SWB		$14,940.00
XL	SuperCab	2WD			$15,785.00
Edge Plus Regular Cab					$15,905.00
XL	Super Cab	2WD	3.0 Liter V-6 Engine		$16,175.00
XLT	SuperCab	2WD			$16,455.00
XLT	Regular Cab	2WD	LWB (Long Wide Bed)	3.0 Liter V-6 Engine	$16,565.00

The XL was the base trim level for the Ranger lineup in 2001 and the XLT was still the deluxe trim level for the Ranger. In between those trim levels Ford introduced a new model they called the "Edge". This youth-oriented model with its aggressive styling cues was clearly aimed at young Ranger buyers with active lifestyles. Ford promoted this new Edge model as having a distinctive appearance at an affordable price for any young, or young minded buyer. The key features that set the Edge apart from the regular Ranger crowd was its monochromatic exterior color treatment in red, white, blue, black, or a striking eye-popping color called Chrome Yellow.

Edge 4x2 and 4x4 vehicles shared the same ride height because they used the same chassis and suspension setups. They also shared the same raised power-dome hood, upper bed trim rails, an AM/FM CD player with a Dolby stereo sound system with 60 watts of power and 4 speakers, a special front fascia bumper cover with special integrated driving lamps, tow hooks, a mesh style grille, Ford's 3.0 Liter EFI V-6 engine and 5-speed manual overdrive transmission, and 5-spoke cast aluminum wheels, and a set of all-terrain radial tires.

All the Ranger trucks this year were built in either the Saint Paul, Minnesota Assembly Plant or the Edison, New Jersey Assembly Plant. Not only were these Rangers sold in the United States and Canada but some of them were also sold in Puerto Rico, US Virgin Islands, South America, Central America, the Caribbean Islands, Guam, New Caledonia, Saipan, Tahiti, and US Samoa. Those 2001 Rangers certainly did get around.

Something much larger and heavier ran into the back end of this Ranger SuperCab pickup truck. At the very least it will need a new or used bed.

At the beginning of the 2001 model year Ford's 2.5 Liter SEFI inline 4-cylinder engine rated at 119 horsepower at 5000 rpm was the base engine found in the XL and XLT Regular Cab 4x2 trucks. Halfway through the model year that engine was replaced by a new smaller 2.3 Liter inline 4-cylinder engine that had a maximum horsepower rating of 135 at 5050 rpm and 153 lb/ft of torque at 3750 rpm.

The base engine for the 4x4 Rangers, SuperCabs,

A chrome plated front bumper, cast aluminum wheels, and chrome plated grille bars gives this Ranger Regular Cab XLT pickup a classier look.

A utility rack mounted in the bed of this Ranger SuperCab XLT model identify it as a work truck rather than a recreational truck.

and the Edge was Ford's 3.0 Liter V-6 engine that carried a maximum horsepower rating of 150 at 5000 rpm with a maximum torque rating of 185 lb/ft at 3750 rpm.

Prior to this model year, Ford's 3.0 Liter V-6 engine was rated as a Flexible Fuel Vehicle… meaning it could be run on 100% unleaded gasoline, E-85 ethanol fuel, or a combination of both. But that FFV rating was dropped for the 2001 Ranger vehicles. However this year both Ranger V-6 engines sold here in the United States and Canada were rated as Low Emission Vehicles.

For years Ranger fans had been calling for more power and some even hoped for a V-8 engine option like Ford's excellent 5.0 Liter (302 CID) V-8. We even saw some Ranger vehicles being equipped with V-8 engines for testing purposes in a shop in California back in the 1990s. So we know that someone at Ford was looking into such an engine option at that time, but those V-8 Ranger fans didn't get their wish for 2001. However, they got the next best thing: a more powerful V-6 engine option for their 2001 Rangers.

This year Ford offered the Ranger buyer the option of a 4.0 Liter V-6 engine that was previously only offered in the Explorer. This 4.0 Liter single overhead cam, fuel injected V-6 engine had a maximum horsepower rating of 207. This was 47 more horsepower than the milder 4.0 Liter V-6 engine that was offered in the Ranger series. Not only did this engine have more horsepower, but at 238 lb/ft of torque, it also had more pulling power.

Under the skin of these new Rangers were a number of changes, including a revised front suspension layout to accommodate the new Ranger engine options and also to provide these trucks with a more refined ride as well as better handling characteristics.

Along with those revised front suspension changes Ford engineers also updated the front end geometry with new stabilizer bar rates, new spring rates, upgraded bushings, and revalved shock absorbers.

The 4x2 Rangers were now fitted with softer rated front coil springs to reduce harshness and vibrations being transmitted into the cab, while 4x4 Rangers and Edge models now used a stronger, heftier torsion bar set up that improved handling without making the ride harsher.

At the rear of these new Rangers, the rear three leaf spring setups were replaced by four leaf spring setups for a smoother ride with the same load ratings.

Other chassis improvements found on the 2001 Rangers included 1 inch larger front brake discs and a new standard four-wheel ABS braking system that was controlled by an electronic brake force distribution system.

The standard transmission once again for the 2001 Rangers was a 5-speed manual overdrive unit. For those Ranger owners who didn't want to use a clutch or shift a transmission, Ford offered two automatic overdrive transmissions at extra cost.

The first automatic transmission that was available with all Ranger engines was Ford's excellent 4R44E, 4-speed unit. The other automatic transmission offered in the Ranger was only available behind the 4.0 Liter V-6 engine and this 5-speed automatic overdrive transmission was referred to as their 5R55E version.

This 5R55E was a smoother shifting unit than the 4R44E because it was controlled by an electronic adaptive shifting computer which governed programming that determined transmission phasing. What that meant in layman's terms was that the computer controlled the firmness of the shifts and which gears were used depending on the driving conditions that the vehicle encountered.

Ford also made a lot of changes to the interiors of these 2001 Rangers to make them quieter and more comfortable. The engineers started with a new firewall design that reduced the amount of engine noise that entered the cab and new engine mounts helped to reduce the amount of harshness that was transmitted into the cab. Another area that was changed in the interior of these trucks was the redesigned weather stripping that was placed around the windows and doors, which did a better job of sealing the cab from the elements as well as cutting down the amount of wind noise that infiltrated the cab.

The 60/40 three passenger split bench seat that was found in most Rangers was now more comfortable and the buyer had his choice of either a Graphite or Medium Tan colored interior trimmings. Ford also took the opportunity while the interior was being freshened up to revise the driver and passenger air bag systems.

Sound systems were also popular options at that time and Ford did their best to supply sound systems that their customers wanted, including an in-dash mounted 6-disc CD changer on top of the regular AM/FM stereo radios offered as standard equipment.

But for those people who wanted one of the best audio systems offered at that time Ford gave them their Ranger "Tremor" aftermarket grade premium audio system as an extra cost option. Actually this system was offered as an option in the latter part of the 2000 model year but Ford didn't really promote it until the 2001 models came out.

This high-powered audio system was rated at 560 watts of power and was only available in SuperCab models because the special sub woofer housing that was built on the floor behind the seat(s) needed the extra room that the SuperCab provided.

The "Tremor" system consisted of an in-dash mounted Pioneer CD/Cassette Player/ and AM/FM stereo radio with selectable audio profiles along with an upgradeable double DIN head unit and a 13-band graphic display. The subwoofer unit built

Rear three quarter view of this Ranger Regular Cab "Edge" pickup in a high profile color shows off its unique upper bed trim rails and color-keyed rear step bumper.

This Ranger SuperCab XLT model is equipped with a canvas covered bed cover. Also note the side step rails, and a mesh style grille treatment.

behind the seat housed a large 10-inch bass speaker and the system also included four other two-way speakers mounted in the doors and the rear passenger areas. Because this system took up so much room in the back of the SuperCab, the standard folding rear jump seats were deleted.

At the end of the 2001 model year Ford had sold 272,460 Rangers, including EVs; a number that was lower than the totals of 1999 and 2000, but still enough to keep the Ranger at the top of the compact truck market.

With all the changes that Ford made to their 2001 Ranger models the 2002 Ford Rangers were introduced without any significant changes. Ford didn't have to make changes for change sake because their Ranger was still the most popular vehicle at the

Though it doesn't look like it at the time this photo was taken this sharp looking 2002 Ranger XLT SuperCab pickup was already thirteen years old.

Here is a sharp looking Ranger SuperCab 4x4 pickup truck that has been dressed up with a two-tone paint job that is augmented with wheel well flares and optional cast wheels.

compact truck end of the market.

Even though automotive writers at the time didn't have any real news to report about the new Rangers their reviews gave the 2002 Ranger lineup high marks for reliability, economy, attractive styling, a solid well-built feel, a very comfortable interior environment, handling, a pleasant driving experience, and the Ranger's overall performance.

The 2002 Ranger lineup was still the same as Ford offered in the previous model year. There was the base model Ranger XL, the deluxe upmarket Ranger XLT, the sporty youth-oriented Ranger Edge, and the Ranger EV electric motor powered vehicle for the environmentally conscious buyer.

Ford also had a new "Ace up their sleeve" in the form of a new optional performance package for their 4x4 SuperCab models this year. This new package, only available on the 4x4 SuperCabs, was called the "FX4 Package" or officially referred to in Ford promotional literature as the "390A Off-Road Vehicle Package". It came with lots of equipment like the following;
• Torsen Limited Slip Rear Differential
• Manual Transfer Case Shift Lever (For manual transmission Equipped Rangers)
• Electronic "Shift-on-the Fly" system (For automatic transmission equipped Rangers)
• 31x10.5 inch BF Goodrich All Terrain T/A KO steel-belted radial tires
• Alcoa Forged 15 inch Aluminum Wheels with an 8 hole design
• Bilstein Shock Absorbers
• Special Sporty Front Bucket Seats
• Stainless Steel Front Tow Hooks
• Black-Colored Rear Tow Hook
• "A Pillar" Assist Handles
• Three Skid Plates
• Black-Colored Wheel Lip Moldings
• Black-Colored Grille
• Black-Colored Front Fascia
• Black-Colored Upper Body Bed Rails
• Ebony or Medium Prairie Tan Colored Interiors
• Color-Keyed Front Rubber Floor Mats
• 4.0 Liter 207 Horsepower Rated SOHC Engine
• Power Equipment Group
• AM/FM Dual Media Audio Head Unit
• Available in The Following Exterior Colors, Black, Bright Red, Silver Frost, or Wedgewood Blue (Medium Prairie Tan interiors only available with exterior colors of Bright Red and Black)

Other changes of note for 2002 Ford Rangers included two new exterior colors for the Ranger XL. These new colors were Harvest Gold Clearcoat Metallic and Silver Frost Clearcoat Metallic. Other XL changes included the choice of two interior colors of Dark Graphite and Medium Prairie Tan.

For Ranger XLT models the new exterior color this year was Dark Highland Green Clearcoat Metallic and there was also a new Sport Bucket Seat Option was available for the XLT SuperCab models. These seats were part of the FX4 package as well. XLT trimmed models also came standard with color-keyed carpeted floor mats, except for the FX4 models which came with color-keyed rubber floor mats.

Ford's "SecuriLock" anti-theft system was standard equipment on all Ford Rangers this year and for those Ford Ranger buyers who liked to load up their trucks with lots of accessories to choose from, Ford added even more optional equipment to their lists.

Another change introduced during the 2002 model year was a new 5-spoke, 16-inch diameter wheel for the Ranger 4x4 Off-Road vehicles, Edge 4x4 vehicles, and SuperCab models with 4.0 Liter V-6 engines.

For audiophiles, Ford offered a new MP3/CD Player Audio Head Unit as an option for the Ranger. Ford also updated their 3.0 Liter V-6 engine this year which gave this engine a little more horsepower while improving its mpg numbers.

This Ranger XLT SuperCab 4x4 pickup in a silver color features optional aluminum wheels and a Flareside bed.

2002 Ranger Model Order Numbers

Regular Cab 4x2 Models	R-10
SuperCab 4x2 Models (2-Door and 4-Door)	R-14
Regular Cab 4x4 Models	R-11
SuperCab 4x4 2-Door Models	R-15
SuperCab 4x4 4-Door Models	R-45

This Ranger Regular Cab long-bed pickup truck is equipped with standard stamped steel wheels and a black pickup cap.

2002 Ranger Interior Trim Colors and Codes

XL	Vinyl Covered 60/40 Split Bench Seat Medium Prairie Tan	HX
XL	Vinyl Covered 60/40 Split Bench Seat Dark Graphite	HT
XL	Cloth-covered 60/40 Split Bench Seat Medium Prairie Tan	EX
XL	Cloth-covered 60/40 Split Bench Seat Dark Graphite	ET
Edge	Cloth-covered 60/40 Split Bench Seat Medium Prairie Tan	EX
Edge	Cloth-covered 60/40 Split Bench Seat Dark Graphite	ET
XLT	Cloth-covered 60/40 Split Bench Seat Medium Prairie Tan	EX
XLT	Cloth-covered 60/40 Split Bench Seat Dark Graphite	ET
XLT	Cloth-covered Sport Bucket Seats Medium Prairie Tan	GX
XLT	Cloth-covered Sport Bucket Seats Dark Graphite	GT
FX4	Cloth-covered Bucket Seats Medium Prairie Tan	SX
FX4	Cloth-covered Bucket Seats Ebony	SB

2002 Ranger Exterior Color Choices and Codes

XL	Harvest Gold Clearcoat Metallic	B2	XLT	Harvest Gold Clearcoat Metallic	B2
	Bright Red Clearcoat	E4		Bright Red Clearcoat	E4
	Silver Frost Clearcoat Metallic	TS		Toreador Red Clearcoat Metallic	FL
	Black Clearcoat	UA		Deep Wedgewood Blue Clearcoat Metallic	LL
	Oxford White Clearcoat	YZ		Dark Highland Green Clearcoat Metallic	PX
				Silver Frost Clearcoat Metallic	TS
Edge	Chrome Yellow Clearcoat	BZ		Black Clearcoat	UA
	Bright Red Clearcoat	E4		Oxford White Clearcoat	YZ
	Bright Island Blue Clearcoat Metallic	LZ			
	Black Clearcoat	UA	FX4	Bright Red Clearcoat	E4
	Oxford White Clearcoat	YZ		Deep Wedgewood Blue Clearcoat Metallic	LL
				Silver Frost Clearcoat Metallic	TS
				Black Clearcoat	UA

At the end of the 2002 model year, over 226,000 Rangers found new homes. That number was down by 45,000 units or so but even with that lower number the Ranger was still the "top dog" in the compact truck field.

For the 2003 model year the Ford Motor Company introduced a new Ranger lineup that was improved and more refined. The Rangers were built in the following four assembly plants; Edison, New Jersey; Norfolk, Virginia; St. Paul, Minnesota; and Oakville, Canada.

All 2003 Rangers featured improved braking systems with larger front disc brake rotors, new front brake calipers, larger rear brake cylinders, and a better 4 wheel ABS control system.

This year the Ranger lineup included the Ranger XL, Ranger Edge, Ranger XLT, and the Ranger Tremor. A model that wasn't around this model year was the Ranger EV electric model. After 1500 or so Ranger EV trucks were built from 1998 through 2002, Ford decided to literally pull the plug on these limited production vehicles. Sorry for the pun, I couldn't help myself.

By the way, out of all those original 1500 Ranger EVs produced, it is estimated that only 200-400 still exist. Every now and then one of them shows up for sale.

Once again the XL models were the base level Rangers. Though these entry level Rangers carried the lowest base prices of the line, they were pretty well equipped with standard equipment like platinum painted bumpers and grille trim, black-colored exterior mirrors, argent-colored styled-steel wheels, tie down hooks in the pickup bed, a floor consolette, and black-colored vinyl floor mats.

Other standard equipment included a color-keyed headliner, Day/Night mirror, color-keyed full width instrument panel pad, 12 volt instrument panel power point, electronic AM/FM radio with clock, vinyl or cloth-covered 60/40 split bench seat with child safety restraints, black-colored steering wheel, color-keyed sun visors with right side vanity mirror, tachometer, driver and passenger side air bags, dome light, Solar Tint glass, dual note horn, 4-pin trailer towing harness, "SecuriLock" anti-theft system, intermittent windshield wipers, an inline 4-cylinder engine, and a 5-speed manual overdrive transmission.

The Edge model was next in line in the Ranger mix. This model came standard with color-keyed bumpers, fog lamps, platinum painted honeycomb styled grille; manual, styled exterior mirrors; bed mounted tie down hooks, front tow hooks, color-keyed wheel lip moldings (Styleside Body Only), 15-inch silver-colored, styled-steel wheels with bright caps, and cast aluminum 16-inch wheels on Edge 4x4 vehicles. Ford also offered a set of 5-spoke cast aluminum 16-inch wheels with gray-colored inserts as an option for Edge 4x4 vehicles.

Other Ranger Edge equipment included black textured vinyl floor coverings, a passenger side mounted grab assist bar, a color-keyed headliner, color-keyed full width instrument panel pad, electronic AM/FM stereo radio with single MP3/CD player with clock and 4 speakers, color-keyed sun visors with right hand vanity mirror, cloth-covered 60/40 split bench seat with child seat restraints, tachometer, air conditioning, dome light with door courtesy switches, dual note horn, "SecuriLock" anti-theft system, multi port fuel injected inline V-6 cylinder engine, and a 5-speed manual overdrive transmission.

The XLT was Ford's upgraded deluxe trim level model. It came with such niceties as chrome plated bumpers and grille trim, contoured platinum-colored mud flaps (4x2) or die cut black-colored mud flaps for 4x4 trucks, black-colored front tow hooks, black bed mounted tie down hooks; 15-inch, silver-painted styled-steel wheels (4x2); 16-inch cast aluminum or 16-inch, 5-spoke cast aluminum wheels (4x4), courtesy lights, upgraded color-keyed door trim panels, floor consolette with dual cup holders, color-keyed floor carpeting, soft contrasting cloth color-keyed headliner, color-keyed full width instrument panel pad, color-keyed "A" and "B" Pillar moldings, 12volt power point port on instrument panel, cloth-covered color-keyed 60/40 split bench seats with child safety seat restraints, color-keyed cloth-covered sun visors with vanity mirrors, and a tachometer.

Other XLT standard equipment included driver side and passenger side air bags, AM/FM stereo radio with CD player and clock, two radio speakers, a 4-pin trailer towing wiring harness, "SecuriLock" anti-theft system, multi port fuel injected engine, and a 5-speed manual overdrive transmission.

For the Ranger buyer who wanted a youth-oriented vehicle with a super stereo audio system, Ford offered him or her the chance to buy a Ranger Tremor model this year.

The Tremor version came with color-keyed front and rear bumpers, front mounted fog lamps, platinum-painted honeycomb grille, manual styled mirrors, die cut black mud flaps, black-colored cargo box tie down hooks, color-keyed wheel lip moldings; 16-inch custom-finished, 5-spoke cast aluminum wheels with argent-colored inserts; and color-keyed door trim panels.

Other Tremor equipment included a black-colored textured floor covering, passenger side mounted assist handle, soft contrasting color headliner, dash mounted 12 volt power point port, color-keyed full width instrument panel pad, Day/Night mirror, cloth-covered 60/40 split bench seat, black steering wheel, color-keyed cloth-covered sun visors with right side vanity mirror, tachometer, air conditioning, "SecuriLock" passive anti-theft system, and a 4 pin trailer towing wiring harness.

The Tremor was a 4x2 SuperCab model that also came with a Pioneer AM/FM stereo radio with dual media head unit with 485 watts of power, an enclosed high output amplifier with 10-inch subwoofer and 4-6x8 inch coaxial speakers. This Tremor 4x2 model also came with P235/70R16 all-season, white-outline-lettered radial tires, a higher output alternator, a white-faced instrument panel gauge cluster, and "Tremor" badging on the tailgate.

Tremor models were available in three monochromatic colors of Chrome Yellow Clearcoat, Sonic Blue Clearcoat Metallic, and Black Clearcoat.

To make the Ranger XLT more appealing, Ford offered buyers a chance to "doll up" this popular trim level by making available an extra cost option Ford called their "Bright Appearance Package". This package was available on 4x2 or 4x4 SuperCab trucks with 2 or 4 doors with a Styleside pickup bed. It included chrome plated driver and passenger side step bars with non-skid surfaces, chrome plated bed rails with platinum-colored end caps, and a chrome plated exhaust pipe tip.

For off-road enthusiasts, Ford offered two "FX4/Off-Road Packages". The first one, under order code 394A, was offered for the XLT 4x4 SuperCab and was outfitted with heavy duty Bulge shocks, skid plates, front tow hooks, all-terrain tires, and a limited slip rear axle. The other off-road package was called the "FX4 Level II (order code 390A) and Ford called this package their "True Off-Road Enthusiast Package that is tearing up the trails". This more serious off-road package included a set of Bilstein shocks, Torsen Limited Slip Axle, B.F. Goodrich T/A KO 31x10.5 inch tires, Alcoa forged aluminum wheels, stainless steer tow hooks, "FX4 Level II" badging, and bold two-tone colored seats in ebony with red or blue color accents. Other FX4 Level II equipment included black-colored wheel lip moldings, 3 skid plates, and a set of deep groove slush floor mats to protect the cab carpeting.

Ford also offered an upgrade package for Edge buyers as well. Ford called this package the "Edge Plus Package" (order code 91G) and it came with a set of P235/75R15 all-terrain, white-outline-lettered tires, machined aluminum wheels, and an in-dash 6-disc CD player.

Another upgraded XLT option was the "XLT Appearance Package" (order number 91A) included a set of P225/70R15 all-season, white-outline-lettered tires wrapped around a set of 7-spoke chrome plated wheels.

And another new for 2003 upgrade package option was the "Ranger Wheels and Tunes Package". This package came with unique machined 16-inch, 5-spoke wheels, Michelin Pilot XGT H4 P235/60R16 tires, AM/FM/Single CD/Mach MP3 Player audio system, or an AM/FM in dash 6-Disc CD Changer. Ford said this "Wheels and Tunes Package was designed to provide an aggressive street-tuned look and stance to the Ranger's tough exterior".

Another new change that debuted with the 2003 Ranger was a new more refined 2.3 Liter inline 4-cylinder engine that was now rated at 143 horsepower at 5250 rpm and had a torque rating of 154 lb/ft at 3750 rpm. Additional changes included the use of thicker glass, new door and "B Pillar" door seals, and a new driveshaft tunnel insulator pad to help cut down on noise, vibrations, and harshness being transmitted into the cab.

To help drivers and passengers get into and out of the 4x4 Ranger cabs Ford installed some chrome plated exterior step bars on the Ranger Edge, 4x4 XL Regular Cab and SuperCabs, and all Ranger XLT models excluding the FX4 Level II models. Ford also made an engine block heater standard equipment on Rangers sold in Alaska, Minnesota, North Dakota, South Dakota, Montana, Wisconsin, and Wyoming.

All 3.0 Liter V-6 engine powered Rangers with automatic transmissions this year were rated as "Flexible Fuel Vehicles" again meaning they could be run on 100% regular unleaded fuel, E85 ethanol, or a combination of both.

Interior Trim Codes and Colors

XL Regular Cab/Super Cab Vinyl Covered 60/40 Split Bench Seat
      Medium Prairie Tan   HX
      Dark Graphite   HT
   Cloth-covered 60/40 Split Bench Seat
      Medium Prairie Tan   JX
      Dark Graphite   JT

Edge Regular Cab and SuperCab Cloth-covered 60/40 Split Bench Seat
      Medium Prairie Tan   JX
      Dark Graphite   JT
   Cloth-covered Sport Bucket Seats
      Medium Prairie Tan   KX
      Dark Graphite   KT

Tremor Cloth-covered 60/40 Split Bench Seat
      Dark Graphite   KT

XLT Regular Cab/SuperCab and FX4 Off-Road 60/40 Split Bench Seat
      Medium Prairie Tan   JX
      Dark Graphite   JT
   Cloth-covered Sport Bucket Seats
      Medium Prairie Tan   KX
      Dark Graphite   KT

FX4 Level II Cloth-covered Sport Bucket Seats
   Ebony /Red   LB
   Ebony /Blue   LB

2003 Ranger Exterior Colors and Codes

XL and Unique Fleet Services
- Arizona Beige Clearcoat Metallic — AQ
- Bright Red Clearcoat — E4
- Silver Frost Clearcoat Metallic — TS
- Black Clearcoat — UA
- Oxford White Clearcoat — YZ

Edge
- Chrome Yellow Clearcoat — BZ
- Bright Red Clearcoat — E4
- Sonic Blue Clearcoat Metallic — SN
- Black Clearcoat — UA
- Oxford White Clearcoat — YZ

Tremor
- Chrome Yellow Clearcoat — BZ
- Sonic Blue Clearcoat Metallic — SN
- Black Clearcoat — UA

XLT and FX4 Off-Road
- Arizona Beige Clearcoat Metallic — AQ
- Bright Red Clearcoat — E4
- Toreador Red Clearcoat Metallic — FL
- Deep Wedgewood Blue Clearcoat Metallic — LL
- Dark Highland Green Clearcoat Metallic — PX
- Sonic Blue Clearcoat Metallic — SN
- Silver Frost Clearcoat Metallic — TS
- Black Clearcoat — UA
- Oxford White Clearcoat — YZ

FX4 Level II
- Bright Red Clearcoat — E4

This Ford ad from 2003 describes all the standard equipment you will find in the special Ranger Tremor model.

Even with all these changes, changes that were supposed to make them more appealing… didn't help to increase the numbers of Ranger trucks sold that model year. Still, at 209,125 total sales, Ford's Ranger was still the top selling truck in the compact truck market.

We don't know what the "YES" sign on this Ranger SuperCab pickup's windshield was meant to signify. In any event, it was a sharp looking pickup that we found sitting on a used car lot.

Chapter 9: America Still Loves The Ranger 2004-2005

Ford introduced their 2004 Ranger lineup by saying ..."The Public has spoken-America loves Ranger. For 17 years in a row, Ford Ranger has outsold the competition by having the right size, the right attitude, the right power, and the right options".

That claim was no idle boast as the subject of that statement was the undisputed sales leader in its class for 17 years in a row. To stay at the top of the pack, the Ford Motor Company introduced an improved Ranger lineup for the 2004 model year.

Those improvements included the raised power-dome hood which was now standard equipment on all 2004 Rangers, new improved bushings in the Ranger's suspension systems, modified shocks and springs for a better ride, and redesigned seats with sculpted bolsters and higher seat backs for a more comfortable seating position. Those seats were now covered in an upgraded material and later on in the model year Ford offered leather covered seats as an extra cost option.

For those audiophiles who wanted a better sound system in their trucks, Ford offered a new Pioneer Sound System with 240 watts of power flowing from an in-dash 6-disc MP3 player through seven speakers.

Ford offered some new exterior colors which were;
Dark Shadow Grey Clearcoat Metallic
Dark Green Satin Clearcoat Metallic
Silver Clearcoat Metallic

The Ranger lineup this year included the same trim levels that were offered in the previous model year. The Ranger XL, the Ranger Edge, the Ranger XLT, the Ranger Tremor, the Ranger FX4, and the Ranger FX4 Level II.

The base model XL came standard with the 2.3

Liter DOHC inline 4-cylinder engine with a 5-speed manual/overdrive transmission, anti-lock 4 wheel disc brakes, front and rear stone guards, a black-faced gauge cluster, cloth upholstery, AM/FM stereo radio with clock, a molded black-colored bar grille, tow hooks; and 15-inch, argent-colored styled-steel wheels.

The Edge models came with most of the standard equipment found on the XL models plus the following items; the 3.0 Liter V-6 engine, fog lamps, power windows, locks, mirrors, and remote keyless entry, an AM/FM/Cassette/CD/MP3 audio system, front cloth-covered sport bucket seats, body-colored bumpers, black honeycomb mesh style grille, 16-inch 5-spoke cast aluminum wheels, white-faced gauges, body-colored wheel lip moldings, gas pressurized shocks, fog lamps, "electric shift-on-the-fly" (4 wheel drive), black textured vinyl floor covering, and air conditioning.

The 2004 Ranger Tremor came with most of the equipment found on the XL and Edge trim levels plus a Pioneer 510 watt sound system with 4 - 6x8 inch coaxial speakers, 5-speed automatic/overdrive transmission, 3.0 Liter overhead valve V-6 engine, speed control with a tilt steering wheel, "Tremor" decal, 16-inch 5-spoke machined aluminum wheels, AM/FM/MP3/6 Disc in-dash CD player, multi channel amplifier, 10-inch sub woofer, air conditioning, 60/40 cloth-covered split bench seat, Power Equipment Group, and side steps for 4x4 models.

The Ranger XLTs were once again Ford's deluxe trimmed models. Ranger XLTs came with a black-colored bar grille, chrome plated bumpers and a chrome plated grille surround trim on 4x4 models and the "XLT Appearance Package", 4x2 Rangers came standard with 15-inch, 7-spoke silver or chrome plated styled-steel wheels, while 4x4 Rangers came equipped with 16-inch, 5-spoke aluminum wheels. 4x4 models also came with fog lamps, side step bars, and Ford's "electric shift-on-the-fly" system.

Regular Cab XLT models came with a 60/40 vinyl covered split bench seat while SuperCab models came standard with a cloth-covered split front bench seat.

All Ranger XLT trucks came with floor carpeting and mats, air conditioning, an AM/FM/CD/Clock combination stereo system, and a black gauge cluster with white accents.

Ranger FX4 Off-Road trucks came with the same standard equipment found on the Ranger XLT models like chrome plated bumpers, air conditioning, floor carpeting with protective mats, and an AM/FM/CD Player/Clock stereo system. Add to that equipment list fog lamps, side step bars, limited slip rear axle, skid plates, premium gas pressurized shocks; 16-inch, 5-spoke cast aluminum wheels; P245/7516 all-terrain, white-letter-outlined tires; leather wrapped steering wheel, 4.0 Liter V-6 engine combined with a 5- speed manual overdrive transmission.

The Ranger FX4/Level II trucks came with all the equipment found on the lower-tiered FX 4 Off-Road Package plus Alcoa forged aluminum wheels, B.F. Goodrich all- terrain tires, Bilstein gas shocks, sport bucket seats with console, a Torsen Limited Slip rear axle, AM/FM/MP3/CD/Cassette Player with clock, 3 skid plates, electric "shift-on-the-fly" 4x4 system, and more.

With a lineup like this it was plain to see that the Ranger offered a variety of vehicles that could appeal to a wide variety of compact truck buyers.

2004 Ranger Exterior Color Choices
Oxford White
Arizona Beige Metallic
Toreador Red Metallic
Bright Red
Dark Green Satin Metallic
Sonic Blue Metallic
Sliver Metallic
Dark Shadow Grey Metallic
Black

Once again the "XLT Appearance Package" option was made available to give the Ranger XLT models a more dressed up exterior image. This package included a front chrome plated grille bar arrangement, chrome plated bumpers and 15-inch, all-season, white-letter-outlined radial tires mounted on chrome plated wheels.

The "Pioneer Sound Package" was a new extra cost option that included a multiple channel amplifier, sub woofer, premium speakers, front bucket seats, a tilt steering wheel with speed control, and an AM/FM 6 disc CD/MP3 stereo system.

Ford's "Bright Appearance Package" also made a return engagement for the Ranger this year. Once again this option included a chrome plated exhaust pipe tip, chrome plated side steps, and chrome plated upper body bed rails.

Here we see a blue colored Ranger SuperCab Sport model with a color-keyed fancy looking pickup bed cap.

Front three-quarter view of a bright red colored Ranger "Edge" SuperCab 4x4 pickup truck with Styleside bed.

2004 Ranger Engine Specifications

2.3 Liter DOHC Inline 4-cylinder
 Aluminum block with aluminum heads, 140 Cubic Inch Displacement
 9.7:1 Compression Ratio, Electronic Sequential Multi Port Fuel Injection System
 Bore x Stroke: 3.44x3.70 inches, Maximum Horsepower Rating:143@5200 RPM
 Maximum Torque Rating: 154 Lb/ft @ 3750 RPM
 Dual Overhead Cam Layout

3.0 Liter Overhead Valve V-6
 182 Cubic Inch Displacement, Iron block with iron heads
 Bore x Stroke: 3.50x3.14 inches, 9.6:1 Compression Ratio
 Electronic Sequential Multi Port Fuel Injection System
 Overhead Valve Layout with 2 valves per cylinder
 Maximum Horsepower Rating: 154 @ 5200 RPM
 Maximum Torque Rating: 180 Lb/ft @ 3900 RPM

4.0 Liter SOHC V-6
 Single Overhead Cam layout
 Iron block with aluminum heads
 Bore x Stroke: 3.95x3.32 inches
 245 Cubic Inch Displacement
 9.7:1 Compression Ratio
 Electronic Sequential Multi Port Fuel Injection System
 Maximum Horsepower Rating: 207 @ 5250 RPM
 Maximum Torque Rating: 238 Lb/ft @ 3000 RPM

2004 Ranger Transmissions with Ratios

5-speed Manual with Overdrive For 2.3 Liter I-4 and 3.0 Liter V-6 Engines
 1st Gear-3.72:1
 2nd Gear- 2.20:1
 3rd Gear- 1.50:1
 4th Gear- 1.00:1
 5th Gear- 0.79:1

5-speed Manual with Overdrive 4.0 Liter V-6 Engine
 1st Gear- 3.40:1
 2nd Gear- 2.05:1
 3rd Gear- 1.50:1
 4th Gear- 1.00:1
 5th Gear- 0.79:1

5-speed Automatic with Overdrive (All Engines)
 1st Gear- 2.47:1
 2nd Gear- 1.85:1
 3rd Gear- 1.47:1
 4th Gear- 1.00:1
 5th Gear- 0.75:1

Rear Axle Ratios Per Engine And Manual Transmissions

	2.3 Liter 4	3.0 Liter V-6	4.0 Liter V-6
4x2	3.73:1, 4.10:1	3.73:1, 4.10:1	3.55:1
4x4		3.73:1, 4.10:1	3.73:1, 4.10:1

Rear Axle Ratios Per Engine with Automatic Transmissions

	2.3 Liter 4	3.0 Liter V-6	4.0 Liter V-6
4x2	4.10:1	3.73:1, 4.10:1	3.55:1
4x4		3.73:1, 4.10:1	3.73:1, 4.10:1

This black-colored Ranger SuperCab XLT 4x4 pickup is equipped with Ford's "FX-4 Level II" off-road package.

When Ford introduced their 2005 Ranger compact trucks, they asked potential buyers if they wanted a truck that wasn't afraid of a challenge. Then they answered their own question by saying the 2005 Ranger was that truck.

In their promotional materials about the 2005 ranger, Ford also touted the fact that their Ranger pickup trucks were the best selling compact trucks for over 17 years. They also mentioned the fact that the R. L. Polk Company had conducted a survey that found in the 2000-2003 model years, the Ranger ranked the highest in customer satisfaction and loyalty in the compact truck field. They also found out that Ranger households had the highest percentage of owners who said they would buy or lease another Ranger truck.

The 2005 Ranger model lineup included the same models that were available in the 2004 model year except for the Ranger Tremor. The trim levels returning for 2005 included the XL, XLT, Edge, FX4 Off-Road, and the FX4 Level II.

The XL and XLT Rangers came standard with the DOHC 2.3 Liter Inline 4-cylinder engine while the Edge models came standard with Ford's 3.0 Liter V-6 that now carried a maximum horsepower rating of 148 @ 4900 RPM, and a maximum torque rating of 180 Lb/ft at 3950 RPM.

Base level Ranger XL s came with the 2.3 Liter

A set of custom dark colored taillight lenses adorn the back end of this Ranger XLT 4x2 SuperCab pickup with Styleside bed.

fuel injected 4-cylinder engine with a 5-speed manual/overdrive transmission, steel wheels with P225/70R15 black wall radial tires, black-colored front and rear bumpers, black-colored grille and trim, black -colored manual exterior mirrors, manual windows, a black-colored steering wheel, and a vinyl covered 60/40 split bench seat on Regular Cab versions, and a cloth-covered 60/40 split bench seat on SuperCab trucks.

The Ranger XLT trimmed trucks also came standard with the 2.3 Liter fuel injected inline 4-cylinder engine with a 5-speed manual/overdrive transmission. Other standard XLT equipment

Pickup bed caps like the one shown on this Ranger SuperCab truck were a very popular accessory during the time these trucks were in production.

As you can see by the number of SuperCab photos that we have used in this book these were very popular trucks.

included color-keyed carpeting and floor mats, AM/FM stereo radio with single CD player, cloth-covered 60/40 split bench seats, rear jump seats on the SuperCab models, black-colored front tow hooks, chrome plated bumpers and chrome plated grille surround trim on 4x4 trucks, black-colored exterior mirrors, and black-colored honeycomb mesh grilles on 4x4 models, and front mounted fog lamps on 4x4 models also.

Ranger Edge models this year came standard with Ford's 3.0 Liter fuel injected V-6 engines and 5-speed manual overdrive transmissions. Other Edge standard equipment included body-colored bumpers and grille surround trim, a black honeycomb mesh grille, body-colored wheel lip moldings, black textured vinyl floor covering, 7-spoke steel wheels with P235/75R15 all-terrain, white-letter-outlined radial tires on 4x2 trucks and 7- spoke, chrome cladded steel wheels inside a set of P235/75R15 all-terrain, white-letter-outlined tires on 4x4 Regular Cab Models. SuperCab Edge trucks came standard with 5- spoke wheels and a set of P255/70R16 all-terrain, white-outline-lettered radial tires on 4x4 models along with a set of fog lamps, and an AM/FM/Cassette Player/ Single CD Player with MP3 capability.

FX4 Off-Road Rangers this year came with a 4.0 Liter SOHC V-6 engine and a 5-speed manual overdrive transmission as standard equipment. Other Ranger FX4 Off-Road standard equipment included a limited slip rear axle, black-colored honeycomb

94

mesh style grille, chrome plated grille surround trim, black wheel lip moldings, 2 skid plates, premium gas pressurized shocks, black side step bars; 5-spoke cast aluminum wheels with P255/70R16 all-terrain, white-outline-lettered radial tires, AM/FM stereo radio with single disc CD player and clock, color-keyed carpet and floor mats, leather wrapped tilt steering wheel with speed control, Power Equipment Group, rear vinyl covered jump seats and a sliding rear window.

The FX4 Level II Rangers came with the same standard equipment as the FX4 Off-Road trucks plus silver-toned front tow hooks, black-colored rear tow hook, 3 skid plates, Bilstein gas pressurized shocks, B.F. Goodrich Radial T/A tires in a 31x10.5 inch size, front cloth-covered sport bucket seats, AM/FM stereo radio with single disc CD player with MP3 capability, and black-colored slush type floor mats.

The Ranger Tremor was no longer a separate model for 2005 but for those who still wanted a truck with "Tremor" equipment Ford offered the "Tremor Package" for the Rangers this year.

This optional package still included a 510 watt Pioneer AM/FM stereo radio with an in-dash 6-disc CD changer with MP3 capability, four - 6x6 inch coaxial speakers, and a rear cab mounted enclosure box for a 10-inch sub woofer and multi channel amplifier. Other "Tremor Package" equipment included a set of machine cast aluminum wheels wrapped up in a set of P235/70R16 all-season radial tires with white-outlined-letter accents on 4x2 models while the 4x4 "Tremor" equipped trucks got a set of P255/70R16 owl radial tires, and a "Tremor" decal.

Once again Ford offered a number of special equipment packages to really dress up the appearance of these Rangers to make them more appealing.

Besides the already mentioned "Tremor Package" there was the "Bright Appearance Package" which included a set of chrome plated upper bed rails with black-colored end caps, side step bars with non-skid strips, and a chrome plated exhaust pipe tip.

This "Bright Appearance Package" when combined with the XLT 4x2 "Appearance Package" added chrome plated bumpers, chrome plated grille surround trim, 7-spoke chrome clad steel wheels or a set of optional machined "split-spoke" cast aluminum wheels and P225/70R15 all-season owl tires.

In years past an ashtray and cigarette lighter were usually part of the standard equipment found on

Another photo of a working Ranger truck, this black Super-Cab Ranger XLT has been fitted with a bed mounted ladder rack.

"FX4 Level II" rear quarter decal placed on Ranger trucks with this top off-road option.

these trucks but in 2005 if you wanted your truck equipped with these items you had to order the "Smoker's Package".

For SuperCab buyers who wanted to dress up their trucks they could order the "Bright Trim Group" which came with chrome plated side steps with non-skid strips, and a chrome plated exhaust pipe tip.

Another extra cost package that Ford offered to Ranger buyers this year was called the "BFT Component Package". This package came with a limited slip rear axle, "Payload Package #2", and the Class III Trailer Towing Package. In order to get this "BFT Component Package" the buyer was required to order a V-6 engine. This package was not available on the Edge 4x2 trucks.

The "Payload Package #2" optional package came standard with heavier duty rear springs and premium gas pressurized shocks on 4x4 trucks while 4x2 models, except for the Edge 4x2, came with heavier duty rear leaf springs and heavy duty shocks.

The "Power Equipment Group", which was standard equipment on the FX4 Off-Road and the FX4 Level II equipped Rangers and optional on Edge and XLT models, included power windows, power door locks, side view exterior mirrors, and a remote keyless entry system.

2005 Ranger Exterior Color Choices
Oxford White
Arizona Beige Metallic
Bright Red
Toreador Red Metallic
Dark Green Satin Metallic
Sonic Blue Metallic
Black
Dark Shadow Grey Metallic
Silver Metallic

2005 Ranger Interior Color Choices
Medium Pebble Tan Cloth
Medium Dark Flint (Grey) Cloth
Medium Pebble Tan Vinyl
Medium Dark Flint (Grey) Vinyl
Ebony with Red Cloth
Ebony with Blue Cloth
Medium Pebble Tan Leather
Medium Dark Flint (Grey) Leather
Solid Ebony Black Leather

Here we see a Ranger SuperCab "Edge" pickup with five spoke aluminum wheels which really dress up the exterior looks of this truck.

Rear three quarter view of a Ranger XLT SuperCab 4x4 pickup with bright 8 hole aluminum wheels and fender flares.

We found this ten year old Ranger XLT SuperCab pickup on a used truck lot in 2015 for a price of $7945.00.

This dark metallic grey-colored Ranger SuperCab pickup is still a sharp looking truck that the owner bought back in 2005.

A Ranger XLT 4x4 SuperCab pickup is what we have here. Note the side steps and a set of five-spoke wheels that have been installed on this truck.

We found this work truck Ranger Regular Cab XLT at a business on historic Route 66 in Eastern New Mexico.

05 RANGER
BUILT Ford TOUGH

A tough looking Ranger SuperCab 4x4 "Edge" truck sits on the cover of the 2005 Ranger showroom catalog that was given out to potential buyers.

For business owners, the base model Ranger Regular Cab pickup with a 6-foot Styleside bed made for a perfect run-around-town delivery or service truck.

Chapter 10: 2006-2009 Bolder And More Aggressive Looking Rangers

In the mid-2000s the automotive press started criticizing the Ford Motor Company because their Ranger trucks were starting to look a "little long in the tooth". It was true at that time that the overall look of the Ranger hadn't been updated in quite awhile and the competition now had fresher looking compact trucks .

From Ford's point of view, they didn't feel the need to make drastic changes to the look of their Rangers because, though the styling cues were starting to look a little dated, the Ranger was still the top-selling vehicle in the compact truck end of the market. Another reason that Ford didn't feel the need for a major styling makeover for the Ranger was the fact that the compact truck market was starting to shrink. With customer demand expected to decrease further in the coming years, the company was reluctant to spend a large amount of development money to do a complete restyling job on the Ranger to make it look more contemporary.

The company knew they needed to make some changes to keep the Ranger at the top of its game so they turned to their styling department to freshen up the looks of the truck. What the stylists came up with was a Ranger that had a bolder, more aggressive, athletic look meant to appeal to a wider audience.

Starting at the front of the vehicle where most of the new key design elements were found, the designers incorporated a new "nostril" styled grille that copied the grille design found on the larger, full-sized Ford pickup trucks. They also redesigned the front valance panel, which now had a larger intake opening giving these trucks a meaner, more aggressive look. Another change seen on the front end of these new 2006 Rangers was the use of parking lamps with clear lenses.

The restyled grille on these trucks now had some horizontal bar inserts and the addition of a large

9-inch Ford oval emblem mounted in the center of the grille. At the rear of these restyled Ranger pickups, Ford stylists changed the tail lamp design and also added another 9-inch Ford oval emblem in the middle of the tailgate.

They also came up with a new two-tone exterior paint treatment for the FX4 Off-Road models that gave these trucks a more "macho" appearance.

The 2006 Ranger lineup included a wide range of vehicles with base prices that ranged from a low of $14,450.00 for an XL trimmed model all the way up to $27,000.00 for an FX4 Level II truck.

These new Rangers were still offered in 4x2 or 4x4 chassis layouts, with short and long beds, Regular Cab models, as well as 2 and 4-Door SuperCab models. All told, with the different trim levels and variants, there were twenty-six Ranger combinations. Those twenty-six Rangers were available in six trim levels: the XL, STX, XLT, Sport, FX4 Off-Road, and the FX4 Level II.

The base level XL models came with the following standard equipment, black-colored bumpers; argent-colored, styled-steel 15 inch wheels, black-colored grille trim, black-colored manual exterior mirrors, interval wipers, and black vinyl flooring. Other standard equipment included a Rallye Gauge Cluster, AM/FM radio with clock, a vinyl covered 60/40 split bench seat in Regular Cab models, and a cloth-covered 60/40 split bench seat in the SuperCab. These XL models also came standard with a 2.3 Liter inline fuel injected 4-cylinder engine and a 5-peed manual/overdrive transmission.

The Ranger XLT trimmed trucks came with a set of chrome plated bumpers and chrome plated grille surround trim pieces. The XLT trim level also added a set of bright metal styled exterior mirrors plus a set of 15-inch, 7-spoke wheels for 4x2 models and a set of 16-inch cast aluminum wheels on 4x4 equipped models. Ranger XLT s also came with carpeted floors with protective mats, and an AM/FM stereo radio with a single CD player.

The Edge model was now gone but it was replaced by the Ranger Sport which basically came equipped with the same pieces found on the Edge model from the previous model year.

The Sport model equipment list included a set of front tow hooks, body-colored bumpers and grille surround trim, and body-colored wheel lip moldings. The Sport models also came with a fancier

You won't find this modified Ranger Regular Cab truck running on the street. It is a purpose built, off-road desert racing machine.

looking honeycomb mesh grille, three wheel choices, textured vinyl flooring in the cab, air conditioning, white-faced Rallye Gauges, an AM/FM Stereo with Cassette/CD Player/ MP3 capability, and a cloth-covered 60/40 split bench seat.

Once again the FX4 Off-Road trim level package was only available on SuperCab models. These models came with a limited slip rear axle, black-colored wheel lip moldings, skid plates, premium gas shocks, side step bars; 16-inch, 5-spoke aluminum wheels mounted in a set of all-terrain tires; rear jump seats, leather wrapped steering wheel with speed control, power windows, power door locks, remote keyless entry, and a tailgate top lip protector. Other equipment included a set of fog lights, and a 4.0 Liter V-6 engine.

The ultimate Ranger for those buyers who liked to blaze new trails both on and off road was still the FX4 Level II. These trucks came with a tough Torsen limited slip rear axle, bright metal front tow hooks, forged Alcoa aluminum 16-inch wheels wrapped inside a set of 31x10.5 inch B.F. Goodrich Radial T/A KO all terrain tires, three skid plates, Bilstein gas pressurized shocks, fog lights, unique two-toned sport bucket seats, a chrome plated shifter on manual transmission equipped trucks, a 4.0 Liter fuel injected V-6 engine, an AM/FM Stereo Radio with Cassette/ CD Player/ and MP3 capability, and a special two-tone exterior paint treatment.

And then there was the Ranger STX model, a model that hadn't been in the Ranger lineup since 1997. This new STX trim level Ranger came with fog lights, a 3.0 Liter V-6 fuel injected engine, AM/FM

This Ranger Regular Cab pickup truck has been fitted with a utility pickup cap and a ladder rack.

In late model pickup trucks like this Ranger SuperCab model the color white was very popular

radio, bucket seats, aluminum wheels, and a set of P225/70R15 white-letter-outlined tires, and all the equipment that came standard with the XL models.

Other 2006 changes of note for the Rangers included an expanded "Bright Appearance Package" option and a new "Bright Trim Package". Other option packages for the 2006 Rangers included the "Tremor Package", "Class III Towing Package", "Chrome Trim Package", "Power Equipment Group", "Privacy Glass", bed extenders, improved automatic overdrive transmissions, and more.

Ford also offered three new clearcoat exterior color choices for the 2006 Ranger trucks. These new colors were Torch Red Clearcoat, Red Fire Clearcoat, and Screaming Yellow Clearcoat.

Going into the 2007 model year the Ford Ranger was still getting high marks for its drivability, handling, build quality, styling, and for the fuel economy numbers that the 4-cylinder Ranger engine could achieve. At 29 mpg on the highway, the Ranger 4-cylinder Regular Cab XL was the most fuel-efficient pickup truck sold in the United States and at a base price of $14,575.00 it was probably the least expensive pickup truck sold here.

The 2007 Ranger lineup again consisted of the Ranger XL, Ranger STX, Ranger Sport, Ranger XLT, Ranger FX4 Off-Road, and the Ranger FX4/Level II.

The Ranger XL was offered in a Regular Cab or SuperCab versions in either a 4x2 or a 4x4 chassis layout. These trucks came with a six-foot bed as standard equipment but Regular Cab truck buyers could order a seven-foot Styleside bed for a little extra money. The standard engine for these XL trucks was still Ford's 2.3 Liter fuel injected inline 4- cylinder but Ranger XL buyers who wanted more power could order a 3.0 Liter or 4.0 Liter V-6 engines. A 5-speed manual transmission was the standard transmission for these trucks but for an extra $1000 Ford would drop in a 5-speed automatic/overdrive transmission.

The Ranger STX model was a step up from the XL. Like the XL trim level below it, the Ranger STX truck was offered in Regular Cab or SuperCab configurations. Unlike the XL, the Ranger STX came standard with Ford's 3.0 Liter V-6 bolted to a 5-speed manual/overdrive transmission. Like the previous model year the Ranger STX came with a set of fog lights, AM/FM radio, 15-inch aluminum wheels with white-outline-lettered tires, and a set of sporty bucket seats.

Up next was the Ranger Sport, a model that was available in six different combinations. Those combinations included Regular Cab and SuperCab models, with either a short or a long pickup bed, and in either a 4x2 or a 4x4 chassis layout.

The Ranger Sport, as its name suggests, was a sporty version that came with a lot of equipment that was meant to appeal to sporty and or young buyers. With its front mounted tow hooks, body-colored grille surround trim, body-colored bumper covers, body-colored wheel lip moldings, some wheel choices, air conditioning, and an AM/FM stereo radio with CD and MP3 capability, it was the perfect vehicle for a customer who wanted a vehicle with just that type of standard equipment. If Ford didn't offer the Ranger Sport, buyers would probably build something similar with aftermarket parts and equipment.

For those Ranger buyers looking for a truck that

looked like it could handle most challenges, the Ranger SuperCab FX4/Off-Road model was the perfect choice. For the true off-road enthusiast there was the SuperCab FX4/Level II model.

One hi-tech option for 2007 was the addition of a "Sirius Satellite Radio" option for the Rangers. Ford also added a new "Personal Safety System" for all Rangers this year. This safety system was a tire pressure-monitoring feature that let the driver know if one of his tires was under inflated.

For those people who ordered the "Tremor Package" this year there were two 16-inch wheel types to choose from. These same two wheel choices were offered on the Ranger XLT models, the XLT 4x4 trucks, and the Ranger 4x2 trucks as well.

Two new exterior color choices were Vista Blue Clearcoat Metallic, and Pueblo Gold Clearcoat.

When the 2008 trucks were introduced, there were two less trim levels for buyers to choose from. These two missing trim levels were the Ranger STX and the Ranger FX4/Level II models. It was said at that time, that Ford cut these two Ranger models to simplify the lineup. We don't know whether that is true or not, but for most of Ranger history, one of the main reasons that automotive reviewers liked the Ranger was because of the wide range of models available.

If Ford did cut these Ranger models to simplify the lineup they might have done so because sales of compact trucks were on the decline. Between the years of 2006 and 2008 the compact truck market segment was down by almost thirty percent. By this time, the only true compact pickup trucks being sold in North America were the Ford Ranger and the Mazda B Series that was basically the same truck. All the other trucks that were once competition for the Ranger and Mazda B Series had all grown larger and were now closer to being medium-sized pickup trucks.

Some reviewers said that the Ranger looked dated and it was time again for Ford to spend some money in redesigning their Ranger trucks and offering more model choices like a true "crew cab" four door version. Reviewers also noted that the chassis needed an update and it would be nice if Ford offered a V-8 engine option for the Ranger.

About the same time, rumors started floating around the automotive community that Ford's Ranger was on its last legs and that it was time for Ford to put this horse out to pasture. Not all of the reviews in

We found this sharp looking Ranger Regular Cab "Edge" truck parked at a local commuter station lot near the Albuquerque area.

This Ranger Regular Cab Styleside pickup with minor bumper damage was sitting on a used car lot in Albuquerque in 2015.

the automotive press concerning the 2008 Rangers had a negative tone. Some of the reviewers still liked the fact that the Ranger line still covered a wide price range and that the Ranger trucks could be equipped to cover a wide range of jobs. Others liked the 2008 Ranger because it was one of the cheapest and most affordable trucks on the market. The Ranger also got high marks for its interior comforts, its reliability, and its performance: especially those Ranger trucks equipped with the 4.0 Liter V-6 engines. It also scored quite well on the NHTSA crash tests. In these tests the Ranger scored a "5 out of 5 Star" side impact rating and a "5 out of 5 Star" driver crash rating, and a "4 out of 5 Star" passenger survival rating in these tests. The only one of these tests where the Ranger scored less than a "4 Star" rating was in the roll over test where the Ranger only scored a "3 Star" rating. In looking at all these results, the NHTSA proved

The owner of this business has 5 or 6 Ranger Regular Cab pickups like this one in his service business fleet.

This Ranger has been fitted with a hood mounted bug deflector and a custom looking grille.

that Ford did build "tough" compact trucks.

The Ranger lineup for 2008 was led off by the XL models, followed by the Sport, the XLT models, and the FX4/Off-Road.

The XL still came with its black-colored bumpers, grille, and grille trim, vinyl covered seats, 4-cylinder engine, 5- speed manual overdrive transmission, and other basic items found on this lower-tiered trim package.

Ranger XLT models now featured body-colored bumpers, chrome plated grille surround trim, and an upgraded deluxe trimmed interior.

The Ranger Sport was still dressed out in a fashion to appeal to young buyers especially when those young buyers opted for the "Tremor MP3 Audio and Wheel Package". Even without that extra cost option package, the Ranger Sport was still a very attractive package what with its body-colored bumpers, wheel lip moldings, grille trim, and the honeycomb mesh style grille that graced the front ends of these vehicles.

With the FX4/Level II package now gone, Ford toughened up the FX4/Off-Road trim package by installing a set of Rancho heavy-duty, twin-tube premium gas pressurized shocks, heavier duty rear leaf springs, and skid plates. The FX4/Off-Road Rangers also were equipped with the two-toned sport bucket seats that were formerly used on the FX4/Level II Rangers.

The biggest styling change was a redesigned front bumper that was said to improve fuel economy numbers. Reviewers also said they liked the larger redesigned road lamps installed on some of the new Ranger trucks.

Sales of the 2008 Rangers were down by about 6800 units to 65,872 trucks compared to the 72,711 that were sold in the 2007 model year. That 65,872 was also 26,548 units less than what was sold in 2006 when 92,420 of them left the assembly lines at Ford plants.

Though some people thought the Ford Ranger was on its last legs in 2008, the Ranger line returned for the 2009 model year. It came back with fewer model choices and base prices that ranged from a low of $16,400 to a high of $25,800. The trim levels for the 2009 Ranger model year were the same as the 2008 model year with the XL, Sport, XLT, and FX4 Off-Road.

There were only two engines offered this year in the Ranger. The first engine was the 2.3 Liter fuel injected inline 4-cylinder while the other engine was the 4.0 Liter fuel injected V-6. Both engines had been offered in the Ranger for a number of years. Matching those two engines were two transmissions, the standard 5-speed manual/overdrive unit and the optional 5-speed automatic/overdrive unit available at extra cost.

The Ranger XL was still the price leader and came standard with a vinyl covered 60/40 split bench seat, vinyl flooring, an AM/FM radio with clock and two speakers, intermittent wipers, power steering, trailer hitch with wiring, tire pressure monitoring system, cargo area tie down hooks, 15-inch styled-steel wheels, and a black-colored grille and bumpers.

The Ranger XLT s came with most of the equipment found on XL models plus air conditioning, cloth-covered seating surfaces, carpeting with protective floor mats, body-colored bumpers, a chrome plated

A sporty looking "Flareside" pickup bed adorns the back end of this Ranger SuperCab Model.

grille bar and grille surround trim, and an AM/FM stereo radio with four speakers.

Ranger Sport models also came with body-colored bumpers and grille surround trim, a deluxe honeycomb mesh grille, step bars, skid plates, fog lamps, 15-inch aluminum wheels, plus a Sirius Satellite capable AM/FM radio with auxiliary input port, an Electronic Distribution Brake Force system that better controlled brake pressure at all four wheels, and a set of sport bucket seats.

Last but not least was the Ranger FX4/Off-Road trim level that was once again reserved for the Ranger SuperCabs. This model came with skid plates, step plates, fog lights, Rancho off-road premium gas pressurized shocks, 16-inch aluminum wheels with a set of all-terrain tires, a leather wrapped steering wheel with speed control, Electronic Distribution Brake Force system, special sport bucket seats with adjustable lumbar supports in the driver's seat, a 4.0 Liter V-6, 5-speed manual overdrive transmission, and a Sirius Satellite capable AM/FM stereo sound system.

By the end of the 2009 model year it was readily apparent that the days of the Ranger here in North America were numbered. It didn't help matters that sales for the 2009 Rangers dropped by another 10,000 units to close at 55,600.

A couple of Ranger SuperCab XLT 4x4 trucks pose side by side in a local parking lot awaiting some service.

Here we see a red colored Ranger Regular Cab "Edge" pickup with Flareside bed and color-keyed utility style pickup cap.

Another white-colored Ranger SuperCab pickup truck is seen here. These trucks are still a common sight on our highways and byways around this country.

106

In the foreground of this photo we see a Ranger SuperCab pickup with a Styleside bed while in the background sits a Ranger Regular Cab truck with a Flareside bed.

One could say that the exterior color paint on this Ranger Sport is definitely a high profile, eye-catching color.

Badge-Engineering At Its Best

Earlier in this book we discussed Ford's partnering with Mazda to produce Ford's first compact pickup truck called the Courier in 1972, the same year that Mazda entered the compact truck market with a truck of their own.

Twenty-two years later Ford returned the favor by building the Mazda B Series trucks alongside the Rangers in the Twin Cities Assembly Plant in Minnesota and the Edison Plant in New Jersey. Except for some trim differences the Mazda trucks looked exactly like their Ranger brothers. In the automotive world building two vehicles that look the same except for trim differences is referred to as "badge-engineering".

These Mazda compact trucks were billed as the B-2300, B-3000, and the B-4000. The B-2300 used a 2.3 Liter 4-cylinder engine, the B-3000 used a 3.0 Liter V-6 engine, and the B-4000 used a 4.0 Liter V-6 engine. Like the Ranger most of these vehicles were offered in Regular Cab or SuperCab models and in 4x2 or 4x4 chassis layouts.

On September 16, 2009 Mazda announced their B-Series pickups were being dropped from the United States Market due to declining sales. And thus the Ranger/Mazda B Series saga in the United States came to an end.

If the name MAZDA didn't appear on the tailgate of this truck one might think the truck was a Ford Ranger.

107

A lone Ranger SuperCab pickup truck sits in a parking lot amongst a group of Japanese cars.

Chapter 11: 2010-2012 Heading For The Finish Line

It was clear that the North American version of the Ford Ranger was on its last legs. If you thought otherwise all you had to do was look at the 2010 Ranger lineup when it was introduced. When these new Rangers were released, Ford only listed eleven versions when in years past they offered three or even four times that number.

> The following is a list of what was available in 2010 for the Ranger buyer with their list prices;
>
> | | |
> |---|---|
> | XL Regular Cab 4x2 112 inch WB | $17,820.00 |
> | XLT Regular Cab 4x2 112 inch WB | $18,960.00 |
> | XL Regular Cab 4x2 118 inch WB | $19,105.00 * |
> | XL SuperCab 4x2 2 Door | $19,515.00 |
> | XLT SuperCab 4x2 2 Door | $20,580.00 |
> | Sport SuperCab 4x2 2 Door | $22,155.00 |
> | XLT SuperCab 4x2 4 Door | $22,485.00 |
> | Sport SuperCab 4x2 4 Door | $23,220.00 |
> | XL SuperCab 4x4 2 Door | $23,570.00 |
> | XLT SuperCab 4x4 4 Door | $24,905.00 |
> | Sport SuperCab 4x4 4 Door | $25,800.00 |
>
> * 7 Foot Long Bed Only Offered to Fleet Buyers

You will note that Ford's special FX4/Off-Road model was no longer offered which probably didn't please Ranger fans that liked to take their trucks off the road to blaze new trails.

Though we were getting close to the end of the line for the Ranger in North America, Ford made sure that these trucks were well equipped with the following standard equipment:

Trailer Hitch with Wiring Harness
Front Cup Beverage Holders
Air Conditioning
AM/FM Radio
A Full Complement of Gauges (including Oil Pressure, Voltmeter,
Fuel, Tachometer, and etc.)
Low Pressure Tire Monitoring System
Intermittent Wipers
ABS 4 Wheel Braking System
Electronic Stability Control
Traction Control
Occupant Sensing Airbag
Dual Front Impact Air Bags
Dual Front Side Impact Air Bags
5-Speed Manual/Overdrive Transmission
Thrifty DOHC Fuel Injected Inline 4-Cylinder Engine
(XL and XLT Models)
Power Steering

Besides all that listed standard equipment the base level Ranger XL came with black-colored bumpers and grille trim, a vinyl covered 60/40 split bench seat, vinyl floor covering, black-colored exterior mirrors, P225/70R15 black wall radial tires, and a set of argent-colored steel wheels.
For a little more money a Ranger buyer could get into an XLT trimmed model with some deluxe grade amenities like;

Power Exterior Door Mirrors
Tilt Steering Wheel with Speed Control
Remote Keyless Entry
Power Front Windows
Front Mounted Fog Lights
Ignition Immobilizer
AM/FM Stereo with MP3 capability and CD player
4 Stereo Speakers
Cloth-Covered Bench Seat
Front Center Armrest with Storage Compartment
P225/70SR15 white-outline-lettered, All-season Radial Tires
Aluminum Wheels
Carpet Floor Covering
Body-Colored Front and Rear Bumpers
Chrome Plated "H" Bar Grille on 4x2 Models

Look close and you will see that this Ranger SuperCab pickup truck has a lockable bed cover.

109

This late model Ranger Regular Cab pickup truck with Styleside bed sits atop a car hauler heading to a new owner.

The Ranger with the most standard equipment content was the Ranger Sport. This top-of-the-line SuperCab model was available as a two-door or four-door model, also in a 4x2 or a 4x4 chassis layout.

The Ranger Sport came with a lot of the equipment that was formerly found on the FX4/Off-Road and the FX4/Level II models.

Equipment like three skid plates mounted in the front, middle, and at the rear of the vehicle, an AM/FM Stereo Radio with auxiliary audio input jacks making this radio MP3 capable, CD Player, a 4.0 Liter fuel injected V-6 engine, power door mirrors, remote keyless entry, leather wrapped steering wheel with speed control, power front windows, panic alarm, rear jump seats, fog lamps, and a set of sport bucket seats with lumbar support.

Other standard equipment found on the Ranger Sport included aluminum wheels, power-dome hood; P235/75SR15 all-terrain, white-outline-lettered tires or P255/70SR16 all-terrain, white-letter-outlined tires; side step bars, body-colored bumpers, floor console, power door locks, and front valance mounted fog lamps.

The two most important pieces of safety equipment on these new Rangers were the seat mounted side air bags, and the "Advance Trac with RSC" system. This "Roll Stability Control (RSC) system was the first use of this system in a compact pickup truck.

This "Roll Stability Control" system utilized a gyroscopic roll rate sensor that determined the truck's body roll angle and rollover rate. If this unique rollover rate sensor detected a significant roll angle the system applied countermeasures like applying pressure to one or more brakes or reducing engine power to increase rollover resistance. Making these Rangers safer to operate in all situations.

With fewer models to choose from this model year, it was no surprise to see that 2010 Ranger sales were down once again. But this drop in sales wasn't all that bad because it only amounted to about 236 units. In 2009 Ford sold 55,600 Rangers and at the end of the 2010 model year Ford had sold 55,364 Rangers.

When the 2011 Rangers made their debut, there were only nine versions offered to the Ranger buyer: two Regular Cab models, and seven SuperCab models.

Those models included an XL trimmed Regular Cab 4x2 Ranger and an XLT Trimmed Regular Cab 4x2 Ranger. The other models were three Sport trimmed SuperCab models in two and four-door versions, and in 4x2 and 4x4 chassis layouts. Then there was an XL trimmed 2 door 4x2 SuperCab, and three XLT SuperCab models in 2-door and 4-door versions, and an XLT 4x4 SuperCab model.

Base prices for these new 2011 Ranger trucks ranged from a low of $17,187.00 for an XL trimmed Regular Cab model with a 4-cylinder engine and a 5-speed manual/overdrive transmission to $26,00.00 for a Sport trimmed Ranger SuperCab model with four wheel drive and a 4.0 Liter V-6 engine with a 5-speed manual/overdrive transmission.

The only two engines available for the Ranger this

This Ranger SuperCab "Sport" has been equipped with a slide in utility box setup.

Rob Campbell has owned a number of Ford trucks over the years and he tells us his 2011 SuperCab pickup shown here is the best truck he has owned in quite awhile. (Rob Campbell Photo).

You can still find used Ranger pickups like this SuperCab model on used car lots in most big city areas across the USA.

This Ranger XLT silver-colored SuperCab pickup was a recent trade in at a local Ford dealership on a new Ford F-150 truck.

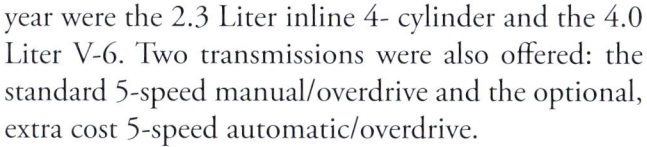

This late model Ford Ranger Regular Cab pickup truck has been fitted with a set of running boards and mud flaps.

year were the 2.3 Liter inline 4-cylinder and the 4.0 Liter V-6. Two transmissions were also offered: the standard 5-speed manual/overdrive and the optional, extra cost 5-speed automatic/overdrive.

2011 Ranger buyers only had a limited number of color choices to choose from this year. Those colors included Torch Red, Dark Shadow Gray Metallic, Black, Oxford White, Silver Metallic, and Vista Blue Metallic. The only interior color available for the Ranger this year was a Medium Dark Flint, otherwise known as dark gray.

Though the Ranger trucks hadn't had a major redesign in some time, it still fared pretty well on the model year sales charts against its main competition: the Nissan Frontier, Toyota Tacoma, and the Chevrolet Colorado. All these models had grown a bit larger and featured designs that were more edgy and more contemporary looking than the Ranger. In spite of the Ranger's dated looks in the 2011 model year, Ford still managed to sell 70,832 of them, a number that was 15,468 units higher than what Ford sold during the 2010 model year.

The 2012 model year would be the last for the Ford Ranger. The base model Ranger for the 2012 model year was the XL trimmed Regular Cab 4x2 with its 2.3 Liter inline 4-cylinder engine. This model was also the mileage champion in the compact truck ranks with its 22 mpg city rating and a highway rating of 27 mpg. That Regular Cab Ranger XL still came with all the same standard equipment that was on it in the previous model year.

The XLT trim level was the next up in the Ranger line for 2012 and it came standard with most of the same equipment that was found on the Ranger XL except for the following upgraded equipment: a cloth-covered 60/40 bench seat, front mounted fog lights, body-colored bumpers, and an AM/FM radio with CD player.

The Ranger Sport was still the top-of-the-line 2012 Ranger model with its sport bucket seats, towing hitch, Rancho gas shocks, body-colored bumpers and body side moldings, a 4.0 Liter V-6 engine, a bed liner, and rear Privacy Glass with a sliding rear window.

In June of 2011 the United Auto Workers Chapter at the Twin Cities Assembly Plant announced that the last North American spec Ranger would be built at this plant in December 2011. Ranger assembly at this plant was originally scheduled to end back in 2007 but steady demand had Ford reconsider that move and had decided years before that the Ranger would stay in production through 2011.

With the news coming out that the North American built Ranger was coming to an end, some Ranger buyers decided that if they wanted a new Ranger truck they better get one now, before they were gone. Because of that interest, Ranger sales for the 2011 model year climbed by over 28% compared to the 2010 model year. But even with that boost in sales, Ford still thought that the compact truck market wasn't big enough to stay in the compact truck market segment; so the company wasn't going to change its mind about pulling the plug on the North American built Ranger trucks at the end of 2011.

The 2012 Ranger model year was a very short one for the North American built Ranger, lasting probably only three months or so. Most of the 2012 models went to fleet buyers like Orkin Pest Control, the company that bought the last 2012 North American Ranger built.

This model, an Oxford White colored 2012 Sport SuperCab model was built on December 16, 2011. This Ranger was not only the last Ranger built at this plant it was also the last vehicle built at this plant. This plant that would be closed soon thereafter, bringing to an end a run that had started almost thirty years before and saw over 7 million North American built Rangers leaving Ford assembly plants Between "Job #1" in 1982 and the last one in 2011.

Rear 3/4 view of a late model Ranger STX Regular Cab model showing off its sporty Flareside bed.

Another Ranger Regular Cab truck awaits its turn to be loaded on the car carrier truck in the background.

This Ranger Regular Cab pickup truck has been equipped with a black composite bed liner to protect the bed from damage.

You don't see too many blue-colored Rangers these days. This particular truck has been dressed up with optional wheels and a decal graphics package.

These two Ranger Regular Cab work trucks sit outside a business waiting for the day to begin. Note the reflective safety tape applied to the tailgate.

You don't see that many black-colored Ranger pickups. This SuperCab XLT 4x4 was found on a Ford dealership used car lot in New Mexico.

The city of Albuquerque, New Mexico has a fleet of small Ranger pickups and most of them are equipped with roof mounted lights and bed mounted toolboxes.

Front 3/4 view of a late Ranger XLT SuperCab pickup truck. Its chrome plated grille and aluminum wheels really dress up the looks of this truck.

Rear 3/4 photo shot of a Ranger Regular Cab model that has been fitted with a black-colored composite bed liner.

Some of the special equipment found on this Ranger SuperCab model include a ladder rack and a lockable bed cover.

The owner of this late model red colored Ranger SuperCab truck has personalized it with bright full length step bars and upper body bed rails.

This photo shows another late model Ranger Regular Cab pickup that has been fitted with a pickup bed cap in a complimentary color.

A set of full length running boards and aluminum wheels really dress up the looks of this Ranger SuperCab Sport model.

A set of black-painted spoke wheels really toughen up the looks of this Ranger SuperCab truck.

This red colored Ranger Regular Cab truck with Styleside bed has been dressed up with a set of optional aluminum wheels.

If you are looking for a tough looking Ranger SuperCab pickup you might want to find one that is equipped like this one.

Even plain wrapper Ranger Regular Cab pickup trucks like this one are still good looking trucks.

Here we see another "Plain Jane" version of a late model Ranger SuperCab pickup truck. Though plain, it is still a good looking truck that most people would be proud to own.

Chapter 12: 1984-1990 Bronco II, Ford's Ranger Based SUV

We can't say for sure when the idea of a downsized Ford sports utility vehicle (SUV) first popped up in a discussion at Ford's World Headquarters building in Dearborn, Michigan. We do think though that it is safe to assume that the subject might have been brought up in the late 1970s when work began on the Ranger project.

Though the full-size Bronco at that time had respectable production figures ranging from 70,000-76,000 units annually, there were probably some Ford truck people who missed having the smaller original Bronco that was in production from 1965 through 1977 in their truck lineup. Those same people probably set the parameters for what a smaller sized SUV should look like.

A year after the Ranger made its debut, the Ford Motor Company released its new downsized sports utility vehicle in March of 1983 as a new 1984 model…a model they branded as the Bronco II.

This new Bronco II was a Ranger-based sport utility vehicle that shared many of its components with the Ranger pickup truck. From the doors forward, the Bronco II used the same hood, doors, grilles, bumpers, etc. As built, the Bronco II was pretty close in size and weight to the original early Broncos.

Like the Ranger pickup truck, the Bronco II featured body-on-frame construction; but unlike the Ranger, the Bronco II was only available in a 4x4 chassis layout.

The Bronco II used the same "Twin Traction Beam" front axle and coil spring independent suspension system setup that was found under Ranger 4x4 vehicles. The rear axle setup found under the Bronco II was similar to the design and concept of the rear axle and suspension setup found under the Ranger as well.

Other items shared between the Ranger and Bronco II were the Warner 1350 2-speed transfer case, 4-speed manual/overdrive transmission; a 2.8 Liter 2V, V-6 engine, and the brake system. The Bronco II also shared the same optional equipment that Ford offered on the Ranger.

This new 1984 Bronco II was offered in three trim levels. The base level model was referred to as the Standard Bronco II, and the other two available Bronco II models were the Bronco II XLT, and the Bronco II XLS.

All three Bronco II models came standard with a 2.8 Liter 2V (170 CID) V-6 engine that had a maximum horsepower rating of 115 at 4600 rpm, a 4-speed manual/overdrive transmission, manual front locking hubs, halogen headlamps, one piece top-hinged tailgate, color-keyed interior carpeting, color-keyed door and trim panels, flow through ventilation system, reclining front bucket seats, a 50/50 fold down rear bench seat, flip down right front passenger seat, Day/Night mirror, color-keyed cloth headliner, inside hood release, color-keyed soft vinyl covered steering wheel, color-keyed sun visors,

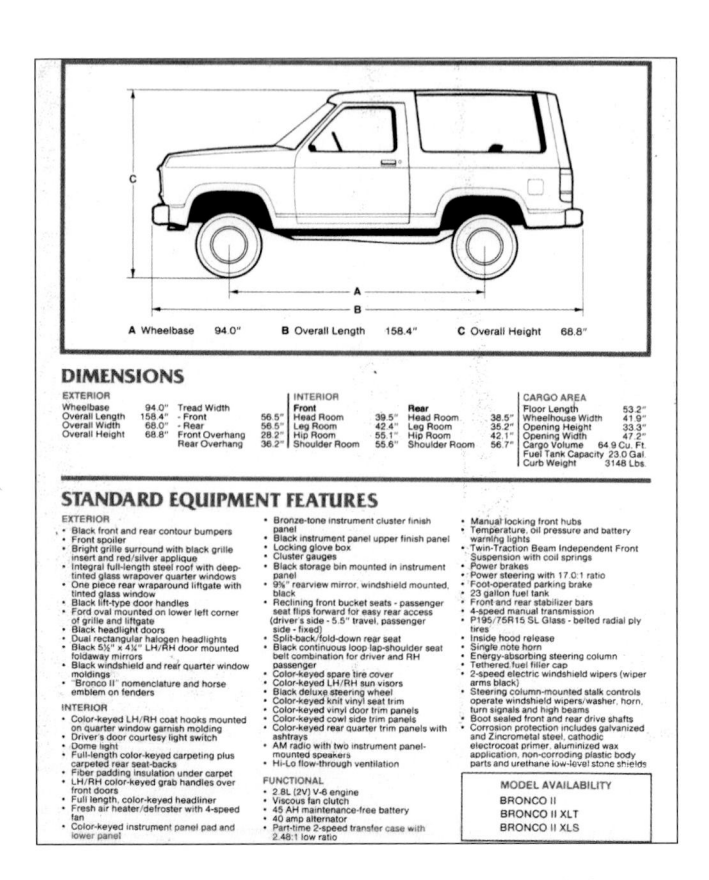

Early Bronco II specification sheet found on the back of a factory brochure.

color-keyed spare tire cover, pewter finished gauge cluster trim, locking glove box, fold away exterior mirrors, deep tined rear windows, and more.

The base model Bronco II came with black-colored bumpers and grille while the Bronco II XLT models came with chrome plated bumpers and grilles, deluxe wheel trim, pivoting vent windows, body side accent striping, and three rear storage compartments.

Interior wise, the Bronco II came with vinyl seat covers and door panels while the upscale Bronco II XLT s came with fancier looking cloth seat covers and cloth-covered door and trim panels.

The Bronco II XLS was the sporty model of the bunch and it came standard with all the pieces found on the base model plus tri-colored special ":XLS" tape exterior striping, rocker panel flare moldings, wheel spats, black-colored grille and trim pieces, black-colored bumpers, and a black-colored sport steering wheel.

1984 Bronco II Exterior Color Choices
Raven Black
Polar White
Light Charcoal Metallic
Midnight Blue Metallic
Medium Desert Tan
Walnut Metallic
Light Desert Tan
Bright Bittersweet
Bright Copper Glow

1984 Bronco II Interior Color Choices
Dark Blue   Canyon Red   Tan

Sometime during the actual 1984 Bronco II model year, Ford joined forces with noted outdoor equipment and casual wear manufacturer Eddie Bauer to offer Bronco II buyers a more upscale, fancier version of the Bronco II.

This new model or trim level was called the Eddie Bauer Bronco II and what made it stand out from the lower tier Bronco II models was its unique two-tone exterior paint schemes of Dark Canyon Red or Medium Blue Metallic colors combined with Ford's Light Desert Tan color.

The Eddie Bauer Bronco II s came with all the standard equipment found on the Bronco II XLT along with a set of cloth-covered Captain's Chairs, "Eddie Bauer" emblems, cast aluminum wheels, and

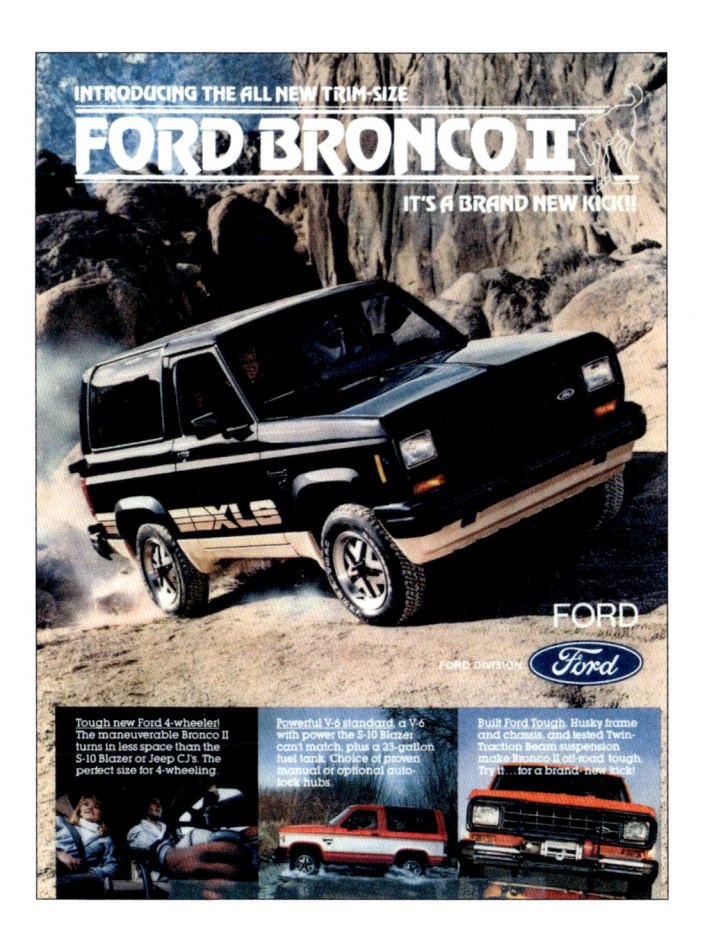

One of the early magazine ads that introduced Ford's Bronco II.

"Ford Care" which was an extended maintenance and limited warranty program of 24 months or 24,000 miles. This "Ford Care" program was on top of the normal warranty program offered on all Ford products at that time.

Ford promoted their new Bronco II models during the calendar years of 1983 and 1984 as being "A Tough New Bronco II...with a Ford Tough Heritage". That promotional slant helped Ford sell 63,178 units in the 1983 calendar year and an additional 98,000 units in the 1984 calendar year.

Introducing a smaller sized SUV was definitely one of Ford's "Better Ideas" for 1984, but Ford wasn't the only car manufacturer to come up with the same idea. Chevrolet also brought a smaller sized SUV to the market about the same time as the Bronco II. Chevrolet called their model the S-10 Blazer and like the Bronco II being based on a Ranger platform, the S-10 Blazer was based on Chevrolet's S-10 compact pickup truck. Unlike the Bronco II that was only available as a 4x4 vehicle in its early days, the S-10 Blazer was available as both a 4x2 and a 4x4 vehicle.

When it was first introduced, the only real

This is the cover photo that was used on a Bronco II booklet that Ford produced for their salesmen. Inside this booklet there is a lot of information about this new model circa 1984.

competition for the Bronco II was the S-10 Blazer; but during the 1984 model year Jeep introduced their new downsized Wagoneer and Cherokee models. The extended 1984 model year turned out to be a good one for the Bronco II and Ford sought to keep up that momentum by making the Bronco II even more appealing to compact truck buyers.

For the Bronco II base model that meant replacing its black-painted bumpers with chrome plated pieces and surrounding its black-colored grille with bright plated trim pieces. This model now came with deep tinted rear windows and sport wheel covers, power assisted steering, a full complement of gauges and an AM radio.

Bronco II XLT improvements included some "XLT" badges, courtesy lights in the ash tray and glove box, a cargo area light, and a "Headlights-on" buzzer. Other XLT standard equipment included an AM/FM stereo radio, cloth-covered seat trims, a leather wrapped steering wheel, vanity mirror mounted on the passenger side sun visor, wood grain trim around the gauges, a dual note horn, bright metal finished low mount Western style swing away mirrors, argent-colored styled-steel wheels, interval wipers, pivoting vent windows, and most of the equipment found on the base model Bronco II.

The Bronco II XLS model was also made to look more appealing this year especially with younger buyers who were looking for a sporty compact sports utility vehicle.

Bronco II XLS models were loaded with standard equipment like an "XLS" interior emblem, cargo area lamp, leather wrapped steering wheel, right hand sun visor mounted vanity mirror, pivoting vent windows, unique rocker panel moldings and wheel spats in Silver Metallic or Medium Desert Tan colors, black-colored bumpers and grille, bright metal grille surround trim pieces, and "XLS" unique exterior tape striping. Other Bronco II XLS equipment included argent-colored styled-steel wheels with bright lug nuts and wheel trim, and a set of reclining front bucket seats.

And last but not least, Ford also made the Eddie Bauer Bronco II models even more appealing than the earlier versions. The 1985 Eddie Bauer Bronco II s contained most of the same equipment found on the Bronco II XLT plus "Eddie Bauer" garment bag, "Eddie Bauer" interior and exterior emblems, "Eddie Bauer" signature premium cloth-covered Captain's Chairs and 50/50 split bench rear seat, floor console, AM/FM stereo radio, pivoting front vent windows, cast aluminum wheels, "Privacy" glass on rear quarter windows, unique "Eddie Bauer" two-tone exterior paint scheme, dual electric horns, and the "Ford Care" extended warranty and maintenance package.

1985 Bronco II Exterior Color Choices
Raven Black
Silver Metallic
Bright Canyon Red
Midnight Blue Metallic
Light Regatta Blue
Dark Canyon Red
Light Desert Tan
Wimbledon White
Dark Charcoal Metallic
Walnut Metallic

1985 Bronco II Interior Color Choices
Regatta Blue   Canyon Red   Tan

Ford also made a 5-speed manual/overdrive transmission standard equipment for all Bronco II models this year and for the first time Ford offered an optional engine for the Bronco II. This engine option was a 2.3 Liter inline 4-cylinder diesel engine.

The calendar year production total of the Bronco II amounted to 111,351 units and calendar year sales were not that far behind with 104,507 Bronco II models finding new homes across the USA and Canada.

For the 1986 model year, Ford made additional improvements across the Bronco II model lineup. One of those improvements involved the installation of a slight larger, more powerful standard engine. This engine was a new 2.9 Liter (179 CID) electronically controlled fuel-injected V-6 engine that had a maximum horsepower rating of 140 at 4600 rpm. What was even better, was that this engine had a torque rating of 170 lb/ft at a low 2600 rpm. In other words, this engine had more than enough power to move any Bronco II out with a minimum of fuss. This would be the only engine used in the Bronco II this year because the diesel engine offered in 1985 was no longer offered.

Also gone for the 1986 model year were the Bronco II XLS models, leaving only the Standard model Bronco II, the Bronco XLT, and the Eddie Bauer Bronco IIs as options.

Another Bronco II change for 1986 was the fact that now one could order any Bronco II model in either a 4x2 or a 4x4 chassis layout; another factor that made the Bronco II appeal to an even wider audience.

Ford promoted the Standard Bronco II as a vehicle that "offers outstanding value in a well-equipped compact utility vehicle". This well-equipped vehicle came with a set of knitted vinyl covered reclining bucket seats along with a full complement of analog gauges, color-keyed carpeted interior, a color-keyed cloth headliner, color-keyed sun visors, halogen headlights, and bright plated contoured bumpers with rub strips.

Other features on the Standard Bronco II were sport wheel covers, tinted glass, a 50/50 fold down rear bench seat, an AM radio, and a black-colored grille with bright surround trim.

For those buyers looking for a fancier vehicle, the Bronco II XLT was the right choice. The front bucket seats in these vehicles were covered in a cloth and vinyl combination as was the 50/50 split rear bench seat. These models also came standard with fancier looking color-keyed cloth-covered door panels with map pockets and lower carpeted areas.

Other XLT equipment for the Bronco II included an AM/FM stereo radio, a new tachometer cluster, driver and passenger side door courtesy lamps, an ashtray light, a lighted glove box, and a light for the engine compartment. Bronco II XLT buyers also got a chrome plated grille, bumpers with rub strips, accent tape stripes, and bright finished low mount Western style mirrors.

This early model Bronco II XLT model is still in pretty good original shape when you consider that this photo was taken in 2015.

Like previous years the Eddie Bauer Bronco II came with all the XLT trim items found on that model plus a unique two-tone paint scheme, accent striping, P205 raised white-letter-outlined, all-season steel-belted radial tires, and a set of Captain's Chairs covered in a rich looking chestnut colored cloth and vinyl material combination.

Other Eddie Bauer themed equipment found on these top-of-the-line models included an AM/FM stereo radio, a tilt steering wheel with speed control, special "Eddie Bauer" garment and tote bags, "Privacy" rear quarter window glass, a passenger side sun visor with vanity mirror attached, and "Ford Care" extended and limited warranty program.

For the 1987 model year the Standard model Bronco II became the Bronco XL. This base model Bronco II came with a set of knitted vinyl covered bucket seats along with a knitted vinyl seat cover on the rear 50/50 split bench seat. Other equipment found on this Bronco II model was an AM radio with digital clock, a trip odometer plus other analog gauges, full length color-keyed carpeting, a color-keyed cloth headliner, halogen headlights, bright finished bumpers with rub strips, sport wheel covers and much more.

1987 Bronco II Base Prices
(4x2 Models) $11,803.00
(4x4 Models) $13,203.00

1987 Bronco Options with Codes and Prices

Code	Description	Prices
572	Air Conditioning (Includes 60 Amp Alternator and Larger Radiator)	$750.00
414	Floor Console/Storage Bin	$174.00
624	Super Engine Cooling	$57.00
213	Electric Shift Touch Drive (4x4)	$104.00
924	"Privacy" Glass (Standard on Eddie Bauer)	$144.00
41H	Engine Block Heater	$33.00
593	Light Group (Standard on Eddie Bauer and XLT) Includes Ash tray light, Glove Box light, "Headlights-on" chime, and Underhood Light	
547	Bright Finished Low Mount Western Style Mirrors (Standard on Eddie Bauer and XLT)	$87.00
952	Paint Deluxe Two-Tone	
	XL Trim	$240.00
	XLT Trim on Special Value Package	$191.00
962	Power Window/Door Locks Group	
	XL Trim with Pivoting Vent Windows	$518.00
	XLT or Eddie Bauer Trims	$310.00
615	Roof Mounted Luggage Rack	$126.00
583	Electronic AM/FM Stereo Radio with Clock and 4 speakers (Standard on Eddie Bauer and XLT)	$93.00
586	Electronic AM/FM Stereo Radio with cassette player, clock, and 4 Speakers	$100.00
91A	Premium Sound System	$121.00
566	Tilt up/ Open Air Roof Panel	$332.00
565	Snow Plow Special Package	
	Includes: Heavy duty 4x4 frame	
	4500 GVWR Package	
	Heavy duty front springs with air bags	
	60 Amp Alternator	
	Must Order Automatic Transmission at extra cost	
	P205 Steel-belted radial tires	
	Without air conditioning	$174.00
	With air conditioning	$118.00
554	Tilt steering wheel with speed control (Standard on Eddie Bauer)	$294.00
51Q	Swing away rear tire carrier and cover	
	Use with P195 all season radial tires	$287.00
	Use with P205 raised white letter all season tires	$317.00
	Use with P205 raised white letter all terrain tires	$343.00
162	Tachometer (XL Trim)	$59.00
646	Cast aluminum wheels (Standard on Eddie Bauer)	
	Use with Preferred Equipment Package 920A	$309.00
	Use without Preferred Equipment Package 920A	$207.00
	Use with Special Value Package	$188.00
647	White sport wheels	
	Use with Preferred Equipment Package 920A	$121.00
	Use with all other Preferred Equipment Packages	$20.00
435	Flip open rear window	$90.00
431	Rear window wiper/washer/defroster	$252.00
	Use with Preferred Equipment Package 933A	$162.00
44T	4-Speed Automatic Overdrive Transmission (Includes transmission cooler)	$735.00
	Limited Slip Rear Axle (Use with 51Q option)	$252.00

Seat Trims		
XL		
A Knitted vinyl covered bucket seats		Standard
J Cloth and vinyl covered bucket seats		$100.00
L Cloth-covered 60/40 split bench front seat		$232.00
XLT		
J Cloth and vinyl covered bucket seats		Standard
B Cloth and vinyl covered Captain's Chairs		$437.00
L Cloth-covered 60/40 split bench front seat		$140.00
Eddie Bauer		
H Cloth and vinyl covered Captain's Chairs		Standard

4x4 Manual Transmission Special Packages	
XL	
920A/602	$184.00
921A/602	$484.00
XLT	
922B/602	$1161.00
930A/602	$2103.00
931A/602	$2070.00

1988 Bronco II Interior Color Trims

Exterior Color	Regatta Blue	Scarlet Red	Chestnut
Raven Black		x	x
Scarlet Red		x	x
Cabernet Red		x	x
Light Chestnut	x	x	x
Colonial White		x	x
Alpine Green Clearcoat Metallic			x
Silver Clearcoat Metallic	x	x	
Shadow Grey Clearcoat Metallic	x	x	x
Bright Regatta Blue Clearcoat Metallic	x		
Deep Shadow Blue Clearcoat Metallic	x		x
Light Chestnut Clearcoat Metallic		x	x
Dark Chestnut Clearcoat Metallic			x

If you wanted a better equipped Bronco II, the Bronco II XLT was once again the one to choose. This upscale model featured a set of cloth and vinyl covered bucket seats and a 50/50 split rear bench seat covered the same way. These models also came with an electronic AM/FM stereo radio, a leather wrapped steering wheel, wood toned accents around the gauges, a tachometer, rear quarter panels with storage compartments, passenger side sun visor with vanity mirror, and a set of door mounted courtesy lights. Speaking of lights these XLT models also came with an ashtray light, a lighted glove box, an under hood light, and a "Headlights-on" warning chime.

Other XLT trim items included some extra sound insulation materials, a chrome plated grille and bumpers with rub strips, deluxe wheel trim, body side and liftgate accent striping, and bright finished low mount Western style mirrors.

This year Ford touted that "driving enjoyment at its best in a compact-sized utility vehicle was what the Eddie Bauer Bronco II was all about".

Once again this top-of-the-line Bronco II model came with its own unique two-tone exterior paint treatment, accent striping, dual reclining Captain's Chairs covered in rich-looking cloth and vinyl fabrics and equipped with built-in lumbar power supports.

Other Eddie Bauer Bronco II standard equipment included all the items found on XLT models plus a tilt steering wheel with speed control, "Eddie Bauer" cloth and leather garment and gear bags, and the "Ford Care" extended maintenance and limited warranty program.

When Ford introduced their 1988 Bronco II models they promoted them by saying "There is no better way to achieve a sense of freedom to choose your own path...and Ford's Bronco II is the fun to drive vehicle to travel in".

The Bronco II lineup for 1988 once again consisted of four models. There was the top-of-the-line Eddie Bauer Bronco II, the Bronco II XLT, the Bronco II XL, and a new a new Bronco II model called the Bronco II XL Sport.

Ford referred to this new Bronco II XL Sport as their "free spirit" Bronco II. That free-spirit attitude was quite evident when you looked at all the standard equipment that Ford put into this model. That standard equipment included a black-painted front bumper, grille, and grille trim, a black-colored grille and brush guard with mounted fog lamps, a black-colored tubular rear bumper, wheel spats, and a unique front valance panel. Other standard Bronco II XL Sport equipment were a tachometer, cast aluminum wheels, and raised white letter steel-belted radial tires.

If you liked all this equipment on this model but preferred a Bronco II XLT model, you could order the same XL Sport trim for your Bronco II XLT by ordering Ford's "Sport Appearance Package" option.

Other changes seen on the 1988 Bronco II models were a new and improved 5-speed manual overdrive transmission with synchronizers on all gears including reverse, a new tip and slide driver's seat to provide easier access to the back seat area, interval wipers on all models, new shift boots, shift knobs, and revised rods for easier shifting of the manual transmission.

1988 Bronco II Exterior Color Choices
Raven Black
Scarlet Red
Cabernet Red
Light Chestnut
Colonial White
Alpine Green Clearcoat Metallic
Silver Clearcoat Metallic
Shadow Grey Clearcoat Metallic
Bright Regatta Blue Clearcoat Metallic
Deep Shadow Blue Clearcoat Metallic
Light Chestnut Clearcoat Metallic
Dark Chestnut Clearcoat Metallic

The 1989 Bronco II models came with a new front end design thanks to a restyling job that featured a new grille, more aerodynamic wraparound headlights with integral turn signals and parking lights, modified front fenders, and bumpers.

Ford referred to this restyling job as…"a new design that is contemporary in style and more aerodynamic in function". Besides changing the front end looks of the new Bronco II s,Ford designers also reworked the instrument panel to make it more driver friendly.

Once again the Bronco II lineup for 1989 consisted of the Bronco II XL, Bronco II XL Sport, the Bronco II XLT, and the Eddie Bauer Bronco II.

The base model Bronco II XL came pretty well equipped this year with a long list of standard equipment like a 2.9 Liter EFI V-6 engine and a 5-speed manual overdrive transmission. Other

Special Value Packages

XLT
922A $1035.00
(4x2) 4-Speed Automatic Overdrive Transmission
   (4x4) 5-Speed Manual Overdrive Transmission
Light Group
Interval wipers
Tachometer
Electronic AM/FM stereo radio with clock
Deluxe wheel trim

930A $1977.00
Air conditioning
   Deluxe two-tone paint
   "Privacy" rear quarter window glass
   Tilt steering wheel with speed control
   Power windows/Door locks
   Rear window wiper/washer/defroster

931A $1944.00
Power windows/Door locks
   Flip open rear window
   P205/75Rx15 inch SL raised white letter all season tires

Eddie Bauer
933A $2613.00
Super engine cooling
   (4x2) 4-Speed Automatic Overdrive Transmission
   (4x4) 5-Speed Manual Overdrive Transmission
Electronic AM/FM stereo radio with clock
Flip open rear window
P205/75R15 raised white-outline-lettered, all-season tires

934A
Electronic AM/FM stereo radio with digital clock and cassette player
   Cloth and vinyl covered Captain's Chairs
   Power windows/Door locks
   Swing away rear tire carrier and cover

standard XL equipment included a set of front high-back bucket seats covered in a knitted vinyl material, a 50/50 split rear bench seat also covered in the same material, a full complement of gauges, a cloth-covered full length headliner, an AM/FM stereo radio with 4 speakers, variable speed interval wipers, black-colored foldaway mirrors, black-colored grille, black-colored bumpers, and sport wheel covers.

The Bronco II XL Sport came with all that equipment plus a redesigned black-colored grille and brush guard with mounted fog lamps, wheel spats, a redesigned black-colored wraparound front bumper, a redesigned black rear tubular bumper and a black-colored rear stone shield, and a special two-tone exterior paint treatment. Other Bronco II XL Sport equipment included deep dish cast aluminum wheels, a flip open rear window, tachometer, and a "Headlights-on" warning chime.

When speaking about the XL Sport model Ford said "The way it is styled and equipped, XL Sport captures in a special way the youthful spirit of fun-to-drive performance engineered into Bronco II".

Up next in the Bronco II lineup was the Bronco II XLT which Ford described as the model that "comes with the features most drivers prefer". They also referred to the Bronco II XLT as the model that best "represents Ford's commitment to offer exceptional value in a sport utility vehicle that is superbly equipped inside and out".

The XLT equipment items that most drivers seemed to prefer were air conditioning, a tachometer, and a leather wrapped steering wheel. Then there were the cloth-covered Captain's Chairs, and 50/50 split rear bench seat, luxury level cloth-covered door panels with their map pockets and lower carpeted areas, a "Headlights-on" warning chime, chrome plated grille and headlight trim pieces, "XLT" badges, and lower body side protection moldings.

If you wanted your Bronco II XLT to look even classier, you could order the Preferred Equipment Package 931A, or "XLT Plus" which included a tilt steering wheel with speed control, "Privacy" rear quarter window glass, flip open rear window, deluxe two-tone exterior paint scheme, and a set of P205/75R15 SL black-sidewall, all-season steel-belted radial tires.

If a sportier looking Bronco II XLT model was more to your liking, you could always order Ford's "Sport Appearance Package" which came with all the special equipment found on the XL Sport.

For those Bronco II buyers who wanted a model that was a notch above an XLT the choice was the Eddie Bauer Bronco II. Once again the Eddie Bauer Bronco II came with all the XLT specific equipment plus the exclusive items found only on the Eddie Bauer models. These items included the "Eddie Bauer" signature premium cloth-covered Captain's Chairs with power lumbar supports and a floor console, P205 white-outline- lettered, all-season steel-belted radial tires, cast aluminum wheels, "Privacy" rear quarter window glass, and an outside swing away spare tire carrier with cover.

Other Eddie Bauer items included its unique exterior two-tone paint scheme with accent striping, an electronic AM/FM stereo radio with cassette player and digital clock, 4 speakers, rear window wiper/washer/defroster system, "Eddie Bauer" signature luggage set, tilt steering wheel with speed control, and the "Ford Care" 24 months or 24,000 mile extended maintenance and limited warranty program.

The 1990 Bronco II models were introduced in October of 1989 and there were no significant changes made to any of the four Bronco II s in the lineup. Most likely that was because these models would only be offered for about four months. The last Bronco II models were built in January of 1990 after which the assembly lines were shut down to begin making preparations for a new sport utility model that would replace the Bronco II.

This new model would make its debut in March of 1990 when it would be credited as a new 1991 model. This model would go on and become one of the most popular sport utility vehicles ever introduced to the automotive market. Of course we are speaking of the Ford Explorer, a model which, like the Bronco II that preceded it, would offer a variety of vehicles to appeal to a wide variety of customers.

Looking back over the seven year history of the Bronco II it was plain to see that this compact sized sports utility vehicle was very popular and proved to Ford and the other automobile manufacturers that there was a definite need for a smaller fun-to-drive vehicle that was just as comfortable traveling on the street or in off-road situations.

During that seven year period Ford produced around 700,000 Bronco II vehicles making this a very profitable endeavor for the Ford Motor Company. Had the Bronco II not been as popular as it turned out to be, the Explorer and other models of this type might never have seen the light of day.

Where Do We Go From Here?

About the same time that Ford was shutting down production of the North American market Ranger at the Twin Cities Assembly Plant they were introducing their new Global Ranger pickup that was built and sold in other markets except for the United States and Canada. This new Global Ranger was bigger and pricier than the Ranger that was previously built and sold here.

During the years that the North American Ranger was built here over 7 million of them left Ford factories and a large number of those Rangers are still on the road today judging by the number of them we see on a daily basis on our highways and byways. Though there hasn't been one of these Rangers built in the United States in over four years these trucks are still very popular today. And a lot of their owners would love to buy another Ranger pickup if they had the chance to do so.

This past summer some of my friends went into one of our local Ford dealerships to buy a new Ranger to replace their old Ranger. When the salesman told them that Ford hadn't built a Ranger pickup since December of 2011 they went down the street and bought a new Toyota Tacoma with all the bells and whistles. It wasn't a cheap pickup by any stretch of the imagination.

When Ford stopped building the Ranger here they hoped that Ranger buyers would just step up and buy a full size F-150 pickup. Some of them probably did but others, like my friends who prefer smaller trucks, just went looking and shopping elsewhere.

For my friends and other Ranger fans we have some news to report, and that good news is that there maybe a "New Ranger in Our Futures".

We have heard rumors that Ford is considering bringing the "Global Ranger" here to the United States and building it at a Ford assembly plant. This new look Ranger won't look like or be as small as the old Ranger we know and love.

Will Ford actually build a new Ranger here and sell it here we can't say for sure. But we think there is a place in the Ford truck lineup for a new Ford Ranger and we would like to see Ford build one and sell it here.

A Tip of the Hat to the Twin Cities Assembly Plant and the Employees That Worked There

Since I wasn't at the plant on December 16, 2011 I can't say for sure but I imagine it was a sad day for those people who were there to finish working on the last Ranger model, a white 2012 Ranger Sport, as it made its way down the line.

Back in the 1950s, as a young boy, one of my favorite things to do was go by the Ford assembly plant that was located in the next town. When the Ford Motor Company decided to close that plant the news was devastating to the local community. Some of the people who worked at that plant were the parents of friends I played with.

For those of you not familiar with the Twin Cities Assembly Plant, it opened back on May 4, 1925 and the first car the plant built was a 1925 Model T Ford Sedan. The last Ford vehicle to leave the plant was that 2012 Ranger Sport.

From 1925 through 2011 it has been reported that almost 9 million Ford vehicles passed through its doors. Besides building Ranger and Mazda B Series pickup trucks, this plant built some 45 or so different Ford vehicles.

This plant was built over a silica mine and that silica was used to make glass that was used in car windshields from 1926 through 1959. On the grounds of the plant there was a hydro-electric plant that produced power for the plant and the electrical energy not used by the plant was put into the grid to help meet the needs of the surrounding area.

During the plant's heyday the facility covered more than 2.2 million square feet spread across 148 acres and it employed about 2000 people. During those heady days, it was reported that this plant could produce around 160,000 vehicles on an annual basis.

December 16, 2011 was definitely a sad day at the Twin Cities Assembly Plant in St. Paul, Minnesota. December 24, 2011 was even a sadder day since that was when the 800 or so employees still working at the plant were given layoff notices because Ford was shutting down the plant permanently…a victim of Ford's "The Way Forward" restructuring program of 2006.

There may be a new Ranger truck coming to the United States but it won't be built at the Twin Cities Assembly Plant in St. Paul because the bricks and mortar that once formed the plant structures have been torn down and carted away.

Paul G. McLaughlin
December 2015